What Sars-Cov2 Taught Me
By: Ini-Herit Shawn P

© Kofi Piesie ReSearch Team. © Same Tree Different Branch

Kofi Piesie/Mossi Warrior Clan
Copyright 2020 by Kofi Piesie Research Team

All rights reserved. No part of this book may be reproduced or transmitted in any form or by any means, electronic or mechanical, including photocopying, recording, or by any information storage and retrieval systems without the written permission of the publisher.

Printed in the United States of America

Table of Contents

Forward - Ankh West

Introduction - Shawn P

Chapter I - Understanding Viruses

Chapter II - Understanding the Immune System

Chapter III - Covid19 (Symptoms/Spread), Transmission, and More

Chapter IV - The Damage Covid-19 Causes

Chapter V - The Evolution of Vaccine Science & The Creation of Covid-19 mRNA Vaccines

Chapter VI - Myths & Pseudoisms Related to Covid-19

Bonus - Origins of Sars CoV2 (Summary)

Bonus - Understanding Breakthrough Infections & Vaccine Effectiveness

Bibliography

Truth is the first victim of war…African Proverb
Dedicated to every person who has lived and died dedicating their lives to the ideas and qualities of Truth in the pursuit to expand, advance, mold, cherish and protect, TRUTH

Acknowledgements

I would like to thank my entire family for allowing me the opportunity to focus on this work along with others. My Teams (Mossi Warrior Clan, Kofi Piesie Research Team) for allowing me to develop into the researcher and mind that I am. Dr. Bonita Coe for sharing the stage with me on more than one occasion, ChillTalk, Ladosha Wright, GayNjorro Skills Academy, Bro. Chavis for introducing me to the Twiv channel, Dr. Maat for allowing me to present a sound argument in favor of vaccine science, Bro Garfield for allowing me to present my first presentation on vaccines on the Dagger Squad Channel, The Scientific Community for allowing us access to the scholarly work that has been done on Covid-19 and vaccines the past year-plus. To all the lives lost to Covid and the families of the people who lost someone to disease, my heart goes out to you and your families. To the future, especially my kids, make sure you remain scientifically literate and learn how to research properly so that one day you can make sound decisions for your family and community! To everyone I have not named ever in a book, this one is for you!!!!! RIP Man, the family misses you cuzzo.

Forward
By: Ankh West

Only a fool test a virus with only the protection from food.

Pseudo Killas

While amid a conversation with my children's mother's side of the family, I quickly realized the lack of scientific literacy on the subject of vaccines. The running theme amongst the family was that the flu vaccine caused you to get the flu. The family members were convinced that the vaccine had the killer flu virus in it. Little did I know that those conversations were preparing me for the pandemic eight months later. Looking back at this situation, the misinformation being spread in the discussion would almost cause a family member to lose their life. The funny part about the misinformation was 15 years prior, I was on the same campaign. How did I get there? It was simple—lack of scientific literacy and confirmation bias. The trap had been set as I left what I called religion. Breaking away from the slave man's information on god was a call we all are familiar with. When I left the confines of all data in one convent place (the bible), I embarked on a new world of endless possibilities. This world had no actual rules on how to get at the truth but would accept anything that was sounding good. We called it sound-right reasoning. If it doesn't sound right, then it stands to reason that it was not. I had gotten trapped in a cult. Without a system of measuring if the information was correct, I was in for a five-year ride of pseudo-isms. Thank goodness there was no pandemic at the time because I would have spread death to my people without even knowing it.

I remember surfing the internet for information on the reliability and safety of vaccines as I awaited the arrival of our newborn. It was a minefield of information pressing me in all directions. How could this be, and why was it so. Fast forward to the summer of 2018 in an Atlanta Hotel. A conversation broke out between me and Divine (suspect) prospect over vaccines. SaNeter was filming the contest as Divine made up all types of misinformation on said subject.

This was the pivotal conversation and one of the reasons this book is being written. From that point, I understood there was a real problem in the black community. The disrespect that followed was incredible. Thank goodness for Shawn the author of this book, who stepped up with one of the first accurate presentations on the topic of vaccines. He quickly realized the problem I had. Research teams were already in place, and all I had to do was steer the conversation to vaccines and viruses. This method has always been effective in past situations. At this point, the topic was on fire, and a literature review followed. Immunology and virology was the most challenging subject I had ever studied. If not for my understanding of biological evolution, the subject would have been incomprehensible. How are we going to make the matter understandable for the community? How are we going to break the fear of modern medicine? This book will provide some answers.

Out of nowhere, a call came to debate the topic against the largest anti-vaccine community in the world that was started in the 1700s in response to West African inoculation by slave colonies in America. That was a clear answer to the smallpox epidemic. Europeans were at odds with each other over using high-level West African science or low-level European folk medicine with Christian prayers. At this point, the Europeans called smallpox the nigga itch and figured we gave it to them. But, unfortunately, this virus gave them no choice. The data that came out of West African inoculation system gave the Europeans their first case studies on immunization. The effectiveness of vaccinations was proven when 300 people were inoculated. Only 6 died at a rate of 2%, and 1,000 people out of 6,000 un-inoculated killed at a rate of 14%.

This debate was epic, filled with over 600 booing white people and about 100 mad at us black people. The stage was set—science versus Pseudo-Science. Little did we know that on Dec 18, 2019, and the unknown virus was in the room. As I stood up and said, "it's not a matter of if another pathogen will come amongst us; it's a matter of when." This book is about that pathogen, the Novel Sars Corona Virus II, and the vaccine that would fight that invisible enemy that has attacked humanity's immune system for the first time in 350,000 years of human history.

Bro Ankh (Godkilla)

Introduction

My journey started before December 2019; it actually began when I became a science enthusiast with aspirations of becoming an avid researcher years prior, but it was not until I listened to a YouTube live stream. I heard people promoting alternative methods for treating illnesses and potential diseases that I started applying proper research methodology to answer a series of questions related to health.

That was not the only reason; the other was that I was on what some called a serious health kick and was looking for ways to lose some weight and regain my health. Like many people who come into what we call consciousness, you suddenly become exposed to a person many people call Dr. Sebi. While researching Dr. Sebi, you'll find out that he promoted an alkaline diet, but he also sold herbal products to help provide additional minerals and vitamins to those who followed his diet; he was heralded as a health guru who cured diseases that have plagued a lot of people. I heard he was involved with several famous people like Michael Jackson, Lisa Left Eye Lopez, and a few others. As I began to look into Dr. Sebi more, I noticed something was missing from his claims. That missing element was his evidence for his claims, and I will admit initially, like many of us, we believe what the masses believe until we began to uncover the truth for ourselves.

So I applied the same research methodology to Dr. Sebi claims as I did with other areas of study, and here lies the issue. His biggest claims are to cure viruses that have plagued mankind for a while now, and I saw that as amazing until I began to look for his published work to support the data. I found absolutely nothing that scientifically supported his claims, and that threw me for a tailspin. I immediately dropped his diet and stopped purchasing his products, and then I began to look into court documents about his claims and practices only to realize he has not cured anyone of what he said he did; however, he helped treat them and provide them with alternatives that prolonged symptoms. At this moment, I had to deal with the reality of understanding viruses, vaccines, and more.

Fast forward back to the YouTube discussion, I was a part of on the channel called Dagger Squad, and the channel owner name is Garfield Reid, a credit specialist from Jamaica who currently resides in the New York area. He would often have all types of people on his channel that pushed certain information, but mainly he taught, debated, or argued about biblical narratives and text. Now I specifically brought up where Garfield resides because the community, we are a part of online or otherwise 80-90% of the information around the community comes from that area. It has been for some time, especially during the era of the Harlem Renaissance.

However, it was April of 2019, and I put together a presentation called 'The Cure' In this presentation, I discussed the history of vaccines, vaccine courts, and vaccine benefits. In that same month, Brother Ankh was a part of several conversations debunking vaccine myths and more. The conversation spilled over to other networks, and by the end of 2019, Brother Ankh had landed himself in a vaccine debate with the leading anti-vaccine talking heads in America, Robert F. Kennedy and Del Bigtree. During that debate, Brother Ankh asked one honest question: "If not vaccines, then what?" To no avail, we did not get an answer to that question at all.

The saddest part about that debate was the fact that Samoa was experiencing a huge MMR outbreak and here we have pushing people to refuse a preventative measure that has worked for hundreds of years. Prior to the Samoa outbreak, Brooklyn, NY, was experiencing issues with a small Jewish community who had seem to be experiencing its own outbreak due to someone who had traveled outside the country returned. Within this moment, everything I have been learning while listening to this debate made me realize that people lack proper research methodology and did not fully understand the immune system.

However, that was not all I noticed in this debate I realized people had been misled by pseudoscientific claims from the anti-vaccine community that has yet to prove its case using sound data but only claims when asked for scientific evidence to prove its case; the argument stayed speculative. The moment Ankh asked Del Bigtree what he would do as a solution for his child against a virus, he stated he would expose his child to the virus and let nature take its course. That response settled the debate for me and pushed me to start understanding the field of virology and immunology right after. Not a few weeks later, Sars Cov2 Coronavirus 19 began to make national news.

I began learning about this particular Coronavirus, understanding that the threat of a pandemic was certain. I read a few open access articles, and then I would share what I've read on any social media page. As the days grew and cases emerged, I just kept reading qualified sources, and over time I began to have some kind of command for the information I was reading. I paid close attention to key parts of certain scientific journals. Those parts were the method, the results, and the conclusion specifically. Most of the scientific jargon I read along the way, I would look up the words and try to gain some understanding; however, some of those words are just extraordinary.

Although I had a basic understanding, I wanted to help the layman properly understand viruses, the immune system, and more about Covid-19 and its effects on people. I also wanted to be sure that people understood the evolution of vaccine science due to an influx of social media myths and pseudoisms that controlled the narrative for quite some time while the scientific community just ignored the noise.

I am of the belief that the general public never fully understood what viruses are, how the immune system actually functions, or how vaccine science has evolved. This is partly due to an organized effort by anti-vaxxers throughout history, and that inability to not know has led to one of the most deadly occurrences in human history.

And now you're about to learn more about What Sars-Cov2 Taught Me!

**Understanding Viruses
Chapter I**

What are viruses? "Viruses are microscopic and infectious; they are called obligated parasites of their host cells. Viruses spread from cell to cell via infectious particles called virions, which contain genomes comprising RNA or DNA surrounded by a protective protein coat. Upon particle entry and disassociation in a host cell, the viral genome directs the synthesis of viral components by cellular systems. Progeny virus particles are formed in the infected cell by de novo self-assembly from the newly synthesized components." (Flint, Racaniello, Rall, Hatziioannou, and Skalka 2020:3)

In Vincent Racaniello's lecture on viruses, he stated: "A virus is an infectious, obligate intracellular parasite compromising of genetic material (DNA or RNA), often surrounded by a protein coat, sometimes a membrane. But it was in his breakdown that caught my attention he states, infectious just means it can go from cell to cell or host to host, obligate intracellular means the virus has to get into a cell in order to reproduce, and parasite means one organism taking something from another." (Racaniello 2021)

In short, viruses are microscopic parasites that enter a host cell which it needs in order to survive; all viruses contain nucleic acid that is either DNA or RNA specific because it can't be both, and viruses cannot replicate themselves without a viable host, so if the virus cannot find another host over a period of time that virus would no longer be able to survive. "Viruses are not living things. Viruses are complicated assemblies of molecules, including proteins, nucleic acids, lipids, and carbohydrates, but on their own they can do nothing until they enter a living cell. Without cells, viruses would not be able to multiply. Therefore, viruses are not living things. When a virus encounters a cell, a series of chemical reactions occur that lead to the production of new viruses. These steps are completely passive; they are predefined by the nature of the molecules that comprise the virus particle. Viruses don't actually do anything. Often scientists and non-scientists alike ascribe actions to viruses such as employing, displaying, destroying, evading, exploiting, and so on. These terms are incorrect because viruses are passive, completely at the mercy of their environment." (Racaniello 2004)

It is very important for us to note that: "When a virus is completely assembled and capable of infection, it is known as a virion. According to the authors of "Medical Microbiology 4th Ed." (University of Texas Medical Branch at Galveston, 1996), the structure of a simple virion comprises of an inner nucleic acid core surrounded by an outer casing of proteins known as the capsid. Capsids protect viral nucleic acids from being chewed up and destroyed by special host cell enzymes called nucleases. Some viruses have a second protective layer known as the envelope. This layer is usually derived from the cell membrane of a host; little stolen bits that are modified and repurposed for the virus to use." (Vidyasagar 2021) It is also important to note that viruses have always existed before mankind; we can acknowledge the fact; "Viruses are all around us, comprising an enormous proportion of our environment, in both number and total mass. All living things encounter billions of virus particles every day. For example, they enter our lungs in the 6 liters of air each of us inhales every minute. They enter our digestive systems with the food we eat, and they are transferred to our eyes, mouths, and other points of entry from the surfaces we touch and the people with whom we interact." (Racaniello, Rall, Hatziioannou, & Skalka 2020:3)

As mentioned earlier its two types of viruses that exist RNA viruses and DNA viruses. RNA viruses are: "viruses in which the genetic material is RNA. The RNA may be either double-or single-stranded." (Davis MD. Ph.D 2021) According to Chuan He, a University of Chicago biologist who studies RNA modifications, "RNA in a basic way is the biomolecule that connects DNA and proteins." (Dar 2020) It's also said that RNA carries the genetic instructions for viruses, and biologist would argue that RNA may be the intricate piece in how life started on earth. DNA viruses are: A virus in which the genetic material is DNA rather than RNA. The DNA may be either double or single-stranded. (Davis MD Ph.D 2021) DNA is basically living material that exists in all life forms or organisms.

The difference between RNA and DNA viruses: DNA viruses are usually double-stranded viruses, while most RNA viruses are single-stranded viruses. Although both can be single or double RNA viruses' mutation rate is higher than that of DNA viruses. Also, DNA viruses replicate within the nucleus of the cell, while RNA viruses replicate within the cytoplasm within the cell wall, not anywhere close to the nucleus of the cell. Remember this last point about RNA viruses and where they replicate. In an up-and-coming chapter, this information will be repeated and dealt with as it relates to vaccine science and how certain vaccines work.

Before we go any further, I would like to make sure that we understand what cells are and their importance. What are cells: "Cells are the basic building blocks of all living things. The human body is composed of trillions of cells. They provide structure for the body, take in nutrients from food, convert those nutrients into energy, and carry out specialized functions. Cells also contain the body's hereditary material and can make copies of themselves." (NIH 2021) Humans are made up of tons of cells; some biologists say millions or maybe even trillions of cells that all play an intricate part in our existence.

We will not have a full biology course on cells, but we will name certain parts of cells that are key components related to the topic of conversation. The goal is to provide you with a basic understanding of viruses and how they operate, and how viruses function within a cell. "Inside the cell, different cell types can look wildly different and carry out very different roles within the body. For instance, a sperm cell resembles a tadpole, a female egg cell is spherical, and nerve cells are essentially thin tubes." (Biggers MD. Ph.D 2021)

According to Dr. Biggers,

Nucleus: "The nucleus can be thought of as the cell's headquarters. There is normally one nucleus per cell, but this is not always the case; skeletal muscle cells, for instance, have two. The nucleus contains the majority of the cell's DNA (a small amount is housed in the mitochondria). The nucleus sends out messages to tell the cell to grow, divide, or die. Cytoplasm: The cytoplasm is the interior of the cell that surrounds the nucleus and is around 80 percent water; it includes the organelles and a jelly-like fluid called the cytosol. Many of the important reactions that take place in the cell occur in the cytoplasm. Ribosomes: In the nucleus, DNA is transcribed into RNA (ribonucleic acid), a molecule similar to DNA, which carries the same message. Ribosomes read the RNA and translate it into protein by sticking together amino acids in the order defined by the RNA." (Biggers MD. Ph.D 2021) To add, Ribosomes are what we consider the each one teach one part of the cell. It spends its time learning so that it can help other cells learn from it. Similar to my role in this book to help you learn all about the things we didn't care to know before the start of a global Pandemic.

Now that we know about parts of the cell, we can identify the different types of viruses that attack other cells causing the spread of infection. In the next chapter, we will discuss more about the immune system and how it eventually finds the virus, but for now, we need to know about the different types of viruses and how they trick the cell into thinking it belongs. HIV, Small Pox, and HPV viruses are all examples of DNA viruses. While Coronaviruses, Zika, and Rabies viruses are examples of RNA viruses. "All viruses must produce mRNA that can be translated by cellular ribosomes. The Baltimore classification allows relationships among viruses with RNA or DNA genomes to be determined based on the pathway required for mRNA production." (Flint, Racaniello, Rall, Hatziioannou, and Skalka 2020:3)

"One of the key features of viruses is their reliance on living cells for replication and propagation. On their own, viruses lack the complete machinery necessary for many life-sustaining functions. Infection of a host cell and viral propagation are dependent on the transcription of viral mRNA, and in turn, the translation of viral proteins as well as genome replication. Specifically, viruses depend on host cells for: (1) energy, mainly in the form of nucleoside triphosphates, for polymerization involved in genome and viral protein synthesis; (2) a protein-synthesizing system for synthesis of viral proteins from viral mRNAs (some viruses also require host enzymes for post-translational modification of their proteins; e.g. glycosylation); (3) nucleic acid synthesis, for although some viruses code for an enzyme or enzymes involved in the synthesis of their nucleic acids, they do not usually contribute all the polypeptides involved and are reliant on various host factors; and (4) structural components of the cell, in particular lipid membranes, involved in virus replication." (Rampersad & Tennant 2018)

Overall we have a running theme when it comes to viruses and how they behave among us. Some are deadly, others are helpful, and some actually have no effect on us. However, viruses do have classification because they are grouped based on size, shape, and a host of other technical terms. "The viruses that infect humans are currently grouped into 21 families, reflecting only a small part of the spectrum of the multitude of different viruses whose host ranges extend from vertebrates to protozoa and from plants to fungi to bacteria." (Gelderblom 1996), But our primary focus in this chapter is on how viruses affect humans. Viruses have been around longer than we have and "the majority of human infections were likely acquired from other animals (zoonoses)." (Flint, Racaniello, Rall, Hatziioannou, and Skalka 2020:3)

It wasn't until the recent pandemic that people really paid attention to the word zoonotic. Despite the emergence of other viruses, a more common virus HIV, a DNA virus, started in monkeys and crossed over into humans. The world is currently still in a pandemic regarding this virus, and because of how this virus functions, it has been really tough to find a solution to this particle virus. But this word zoonotic; "Pertaining to a zoonosis: a disease that can be transmitted from animals to people or, more specifically, a disease that normally exists in animals but that can infect humans." (Davis MD Ph.D 2021) However, the word zoonotic has been a topic of conversation lately, especially regarding the origins of Sars-Cov2. Partly because people are very unfamiliar with how we encounter viruses, another reason is that people like to blame humans for creating or replicating new viruses.

"In addition to the Aids pandemic, the highly fatal Ebola hemorrhagic fever, severe acute respiratory syndrome (SARS), and Middle east respiratory syndrome (MERS) are recent examples of viral diseases to emerge from zoonotic infections. The influenza virus H5N1 continues to spread among poultry birds in areas of the Middle East and Asia. The virus is deadly to humans who catch it from infected birds. The frightening possibility that it could gain the ability to spread among humans is a major incentive for monitoring for person to person transmission in case of infection by this and other pathogenic avian influenza viruses. Given the eons over which viruses have had the opportunity to interact with various species, today's, 'natural' host simply be a way station in viral evolution."(Flint, Racaniello, Rall, Hatziioannou, and Skalka 2020:6)

This is why it is vital for everyone to know something about viruses, their origins, and their capabilities among humanity. Viruses are everywhere and infect all living things. As mentioned above, some viruses have benefits while other viruses cause human disease. Another good point that can be made about viruses is that anthropologists and virologists tend to work together in tracking human migrational patterns. "The study of viruses has contributed in a unique way to the field of anthropology. As ancient humans moved from one geographic area to another, the viral strains unique to their original locations came along with them. The presence of such strains can be detected by analysis of viral nucleic acids, proteins, and antibodies from ancient human specimens and in modern populations. Together with archeological information, identification of these virological markers has been used to trace the pathways by which humans came to be inhabit various regions of our planet." (Flint, Racaniello, Rall, Hatziioannou, Skalka 2020:6)

The study of viruses helps solidify answers to questions that scientists have had for some time but could not really come to a scientific consensus on certain questions due to the lack of data that was available until the field evolved. Now we can track new virus origins per-say to a specific location that allows scientists to create those virological markers mentioned above so that scientists can trace the viruses and locate potential virus mutation, which can create new virus variants. "Virus mutations create genetic diversity, which is subject to the opposing actions of selection and random genetic drift, both of which are directly affected by the size of the virus population." (AdiStern & Andino 2016)

"A variant is an alteration in the most common DNA nucleotide sequence. The term variant can be used to describe an alteration that may be benign, pathogenic, or of unknown significance. The term variant is increasingly being used in place of the term mutation." (NIH 2021) According to Vince Racaniello, "A virus variant is an isolate whose genome sequence differs from that of a reference virus. No inference is made about whether the change in genome sequence causes any change in the phenotype of the virus. The meaning of variant has become clouded in the era of whole viral genome sequencing because nearly every isolate may have a slightly different genome sequence." (Racaniello 2021)

Virologists have not settled on this word variant or strained as both are used interchangeably. Variant became very popular during the pandemic; however, this does not mean it's a different virus; it just means the virus can be more transmissible or infectious, maybe even according to science. Science accredits Charles Darwin for natural selection because it played a major role in helping scientists understand adaptation and mutation.

"When an environmental change occurs, species are able to adapt in response due to mutations in their DNA. Although these mutations occur randomly, by chance, some of them make the organism better suited to their new environment. These are known as adaptive mutations." (Enard 2016) Now that we have an understanding of how viruses behave and live among us, some thriving to make us who we are today, and while other viruses are deadly and tend to control the outcome of millions of lives, let's learn more about what Sars-Cov2 taught me.

According to the NIH (National Institution of Health), "Coronaviruses are a large family of viruses that usually cause mild to moderate upper-respiratory tract illnesses, like the common cold. However, three new coronaviruses have emerged from animal reservoirs over the past two decades to cause serious and widespread illness and death. There are hundreds of coronaviruses, most of which circulate among such animals as pigs, camels, bats, and cats. Sometimes those viruses jump to humans—called a spillover event—and can cause disease. Four of the seven known coronaviruses that sicken people cause only mild to moderate disease. Three can cause more serious, even fatal, disease. SARS coronavirus (SARS-CoV) emerged in November 2002 and caused severe acute respiratory syndrome (SARS). That virus disappeared by 2004. The Middle East respiratory syndrome (MERS) is caused by the MERS coronavirus (MERS-CoV). Transmitted from an animal reservoir in camels, MERS was identified in September 2012 and continues to cause sporadic and localized outbreaks. The third novel coronavirus to emerge in this century is called SARS-CoV-2. It causes coronavirus disease 2019 (COVID-19), which emerged from China in December 2019 and was declared a global pandemic by the World Health Organization on March 11, 2020." (NIH 2021) Now that we know that Coronavirus is not just one particular virus, maybe we can be more specific and address the latest Coronavirus by its proper name SARS-CoV2.

"The Coronavirus Study Group (CSG), part of the International Committee on Taxonomy of Viruses (ICTV), was responsible for naming SARS-CoV-2. The CSG is responsible for developing the classification of all viruses and taxa in the coronavirus family. The ICTV was created after the International Congress of Microbiology in Moscow in 1966. The committee's purpose was to develop a universal taxonomic scheme to categorize and classify all viruses that infect animals, plants, fungi, bacteria, and archaea. The ICTV is responsible for regulating the Code of Nomenclature, a set of rules and guidelines for naming organisms, and for approving the creation of virus taxa, which are the orders, families, subfamilies, genera, and species of known viruses. The virus (SARS-CoV-2) and the disease caused by the virus (COVID-19) have different names because each name serves a different purpose. Viruses are named based on their genetic structure to aid the development of diagnostic tests and vaccines. Diseases are named to aid discussion on disease prevention, spread, and treatment." (Zoppi BA 2021)

- Co – corona
- Vi – virus
- D – disease
- 19 – the year in which the outbreak began.

Chapter I

Review

1). What is a virus?

2). What are virions?

3). Are DNA and RNA viruses the same?

4). Obligate intracellular means the virus has to get into a cell in order to reproduce. True or False?

5). How many different types of viruses exist?

6). Can virus origins be traced back to a geographical location?

7). What are cells?

8). Do viruses mutate and adapt?

9). Can viruses jump species and if so, what is it called if they do?

10). Are viruses alive or dead?

11). What are virus variants?

12). _____ protects viral nucleic acids from being chewed up and destroyed by special host cell enzymes called nucleases.

13). The study of viruses has contributed in a unique way to what field of study?

14). All viruses must produce _____ that can be translated by cellular ribosomes.

Chapter I Sources

- Biggers, Alana, and Tim Newman. "The Cell: Types, Functions, and Organelles." Medical News Today, MediLexicon International, 8 Feb. 2018, www.medicalnewstoday.com/articles/320878#division.

- Cann, Alan J. "Virus Mutation." Virus Mutation - an Overview | ScienceDirect Topics, 2012, www.sciencedirect.com/topics/biochemistry-genetics-and-molecular-biology/virus-mutation.

- Control, Centers Disease. "About Variants of the Virus That Causes COVID-19 ." Centers for Disease Control and Prevention, Centers for Disease Control and Prevention, 20 May 2021, www.cdc.gov/coronavirus/2019-ncov/variants/variant.html.

- Davis, Charles Patrick. "Medical Definition of DNA Virus." RxList, RxList, 29 Mar. 2021, www.rxlist.com/dna_virus/definition.htm.

- Davis, Charles Patrick. "Medical Definition of RNA Virus." MedicineNet, MedicineNet, 29 Mar. 2021, www.medicinenet.com/rna_virus/definition.htm.

- Davis, Charles Patrick. "Medical Definition of Zoonotic." MedicineNet, MedicineNet, 29 Mar. 2021, www.medicinenet.com/zoonotic/definition.htm.

- Davis, Charles Patrick. "What's a Virus? Viral Infection Types, Symptoms, Treatment." MedicineNet, MedicineNet, 6 Oct. 2020, www.medicinenet.com/viral_infections_pictures_slideshow/article.htm.

- Dhar, Michael. "What Is RNA?" LiveScience, Purch, 15 Oct. 2020, www.livescience.com/what-is-RNA.html.

- Enard, David, et al. "Viruses Are a Dominant Driver of Protein Adaptation in Mammals." ELife, ELife Sciences Publications, Ltd, 17 May 2016, elifesciences.org/articles/12469#digest.

- Flint, S. Jane, et al. Principles of Virology. Fifth ed., I, American Society for Microbiology, 2020.

- Gelderblom, Hans R. "Structure and Classification of Viruses." Medical Microbiology. 4th Edition., U.S. National Library of Medicine, 1 Jan. 1996, www.ncbi.nlm.nih.gov/books/NBK8174/#:~:text=A%20complete%20virus%20particle%20is,inside%20a%20symmetric%20protein%20capsid.

- Health, National Institute. "Coronaviruses." National Institute of Allergy and Infectious Diseases, U.S. Department of Health and Human Services, 26 Mar. 2021, www.niaid.nih.gov/diseases-conditions/coronaviruses.

- Health, National Institute. "What Is a Cell?: MedlinePlus Genetics." MedlinePlus, U.S. National Library of Medicine, 22 Feb. 2021, medlineplus.gov/genetics/understanding/basics/cell/#:~:text=Cells%20are%20the%20basic%20building%20blocks%20of%20all%20living%20things.&text=Cells%20also%20contain%20the%20body's,certain%20tasks%20within%20the%20cell.

- Health, National Institute. "What Is DNA?: MedlinePlus Genetics." MedlinePlus, U.S. National Library of Medicine, 19 Jan. 2021, medlineplus.gov/genetics/understanding/basics/dna/.

- Kuhn, Jens H., et al. "Virus Nomenclature below the Species Level: a Standardized Nomenclature for Natural Variants of Viruses Assigned to the Family Filoviridae." Archives of Virology, U.S. National Library of Medicine, 23 Sept. 2012,

- www.ncbi.nlm.nih.gov/pmc/articles/PMC3535543/#__ffn_sectitle.

- Racaniello, Vincent. "Vincent Racaniello." Virology Blog, 9 June 2004, www.virology.ws/2004/06/09/are-viruses-living/.

- Rampersad, Sephra, and Paula Tennant. "Replication and Expression Strategies of Viruses." Viruses, U.S. National Library of Medicine, 30 Mar. 2018, www.ncbi.nlm.nih.gov/pmc/articles/PMC7158166/#:~:text=All%20viruses%20must%20therefore%20express,vary%20among%20different%20viruses%20(Fig.

- Stern, Adi, and Raul Andino. "Viral Evolution: It Is All About Mutations." Viral Pathogenesis (Third Edition), Academic Press, 12 Feb. 2016, www.sciencedirect.com/science/article/pii/B9780128009642000173.

- Vidyasagar, Aparna. "What Are Viruses?" LiveScience, Purch, 6 Jan. 2016, www.livescience.com/53272-what-is-a-virus.html.

- Zoppi, Lois. "The Naming System Behind SARS-CoV-2." News, 9 Mar. 2021, www.news-medical.net/health/The-Naming-System-Behind-SARS-CoV-2.aspx.

Understanding The Immune System
Chapter II

As a researcher writing this chapter was one of my toughest challenges, and when I look back to 2019 at my very first lecture regarding vaccines, I was unable to make a few key points that later would evolve over time as I continued to study. Many people who argued against vaccines would offer alternative methods to treat infection, not prevent infection, and that point seems to go unheard by those who felt like herbs could cure and rid the body of any virus. This led me to believe that people did not understand viruses at all, and neither did they understand how the immune system functions. Social media conversations began to be centered on discussion driven by vaccines that would shed light on my belief, further allowing one's ignorance to demonstrate their lack of knowing but also their lack of rejecting the evidence. Most people assume that you can boost your immune system because that is a word that is often used by herbalists and certain online websites.

I've learned to use the word strengthen from my brother Nehisi because, as he states: "boosting could kill you but strengthen the immune system and making it aware as in waking it up would be a more accurate." So, the proper word would be strengthened, and we must look at the immune system as a muscle; it must be trained and strengthen just like a muscle to help protect and defend the body against foreign pathogens that it encounters. "The idea of boosting your immunity is enticing, but the ability to do so has proved elusive for several reasons. The immune system is precisely that — a system, not a single entity. To function well, it requires balance and harmony. There is still much that researchers don't know about the intricacies and interconnectedness of the immune response. For now, there are no scientifically proven direct links between lifestyle and enhanced immune function." (Harvard Health Publishing 2021)

In this chapter, we will set out to properly address the immune system, its role, and more. Why? Because we need to know how the immune system functions, what we can do to help strengthen it properly, and the significant role it plays in our lives from the time we are conceived until the time we transition and leave this earth.

Sooooo……, What is the immune system? In short: "The immune system is the body's defense against infections. The immune system attacks germs and helps keep us healthy." (Hirsch 2019) According to the NIH (National Institute of Health), "A complex network of cells, tissues, organs, and the substances they make that helps the body fight infections and other diseases. The immune system includes white blood cells and organs and tissues of the lymph system, such as the thymus, spleen, tonsils, lymph nodes, lymph vessels, and bone marrow." Dr. Davis defines the immune system as "A complex system that is responsible for distinguishing a person from everything foreign to him or her and for protecting his or her body against infections and foreign substances." (Davis 2021) To the layman the immune system helps the body fight off infections and diseases. Humans possess 3 types of immunity - Innate (Born with it), Adaptive (Develops over time), and Passive Immunity (Borrowed). The immune system takes a while to develop immunity against a foreign pathogen, which can cause serious problems within the body. People whose immune system that has limited ability to fight off infection or disease would be considered as immunocompromised.

"Innate Immunity, or nonspecific, immunity is the defense system with which you were born. It protects you against all antigens. Innate immunity involves barriers that keep harmful materials from entering your body. These barriers form the first line of defense in the immune response. Examples of innate immunity include Cough reflex, Enzymes in tears and skin oils, Mucus, which traps bacteria and small particles, Skin, and Stomach acid." (NIH 2021)

Here are examples of how Innate Immunity helps us in our fight against foreign pathogens: Boogers can be very valuable in fighting pathogens that try to enter through the nose, our skin is the largest organ in our body, so its role is significant to us all, and tears do a great job of washing away potential threats that try to enter through the eyes. That yawn that draws tears early in the am or during the day has its benefits.

"Adaptive Immunity refers to antigen-specific immune response. The adaptive immune response is more complex than the innate. The antigen first must be processed and recognized. Once an antigen has been recognized, the adaptive immune system creates an army of immune cells specifically designed to attack that antigen.

Adaptive immunity also includes a "memory" that makes future responses against a specific antigen more efficient." (Arizone Edu 2000) Adaptive speaks for itself, and we all have been in some situations where we had to adapt to our environment or certain circumstances. This helps us build up a tolerance for things we did not have a tolerance with before, so if we encounter that which we built up a tolerance for we would know how to respond to it and that is how adaptive immunity works.

"Passive Immunity is due to antibodies that are produced in a body other than your own. Infants have passive immunity because they are born with antibodies that are transferred through the placenta from their mother. These antibodies disappear between ages 6 and 12 months. Passive immunization may also be due to injection of antiserum, which contains antibodies that are formed by another person or animal. It provides immediate protection against an antigen but does not provide long-lasting protection. Immune serum globulin (given for hepatitis exposure) and tetanus antitoxin are examples of passive immunization." (NIH 2021)

We have examples of passive immunity at work regarding Sars-CoV2 Cov-19. Doctors have found antibodies in the placenta in babies after they had been born. That is a great example of passive immunity at work, and they have also confirmed that immunity was not long-lasting at some point those children would need to be vaccinated.

Once I began to understand how the immune system behaves and functions, I began to look at viruses differently and question herbs more as not a possible solution to a novel virus. However, I couldn't ignore other parts of the immune system that help join in the fight to prevent foreign pathogens from attacking the body. White blood cells play an important role in the immune system because they remember foreign pathogens (lymphocytes) and also dissolve invaders (phagocytes).

"One type of phagocyte is the neutrophil, which fights bacteria. When someone might have a bacterial infection, doctors could order a blood test to see if it caused the body to have lots of neutrophils. Other types of phagocytes do their own jobs to make sure that the body responds to invaders. The two kinds of lymphocytes are B lymphocytes and T lymphocytes.

Lymphocytes start out in the bone marrow and either stay there and mature into B cells or go to the thymus gland to mature into T cells. B lymphocytes are like the body's military intelligence system — they find their targets and send defenses to lock onto them. T cells are like the soldiers — they destroy the invaders that the intelligence system finds." (Hirsch 2019) I read a recent study in the journal of nature (actually an abstract) that spoke to the Lymphocytes and projecting immunity based on bone marrow. Although the study was inconclusive, it basically fell directly up under the explanation of what B and T lymphocytes represent.

"The immune system needs to be able to tell self from non-self. It does this by detecting proteins that are found on the surface of all cells. It learns to ignore its own or self-proteins at an early stage." (Newman 2018) This is why natural infection can be dangerous because it takes the immune system a while to adapt to a foreign pathogen and create a defense against it. "The capacity of the immune system to recognize self-antigens and accept the presence of normal cells is known as self-tolerance. When self-tolerance fails, it can result in autoimmunity, whereby the immune system may fail to discriminate self from non-self and attack normal cells." (Fuertes MB, Corrales 2014)

"Antigens, small molecules, or peptides capable of eliciting an immune response are key elements in this process of distinguishing self from non-self. Antigens serve as labels that enable the immune system to distinguish between a normal interaction (self) and an encounter with a foreign threat (non-self)." (Warrington R, Watson W, Kim HL, Antonetti FR 2011) I found it interesting that both adaptive and innate immune systems both can tell the difference between self and non-self.

"Innate immune response is rapid, while adaptive immune response is not as immediate but can produce a durable response through the development of memory cells, including memory T cells. Chronic exposure to a non-self-antigen can promote the accumulation of memory T cells. Innate and adaptive immunity is activated through distinct and often complementary mechanisms that deploy different effector cells to attack and destroy abnormal/foreign cells such as cancer." (Ghiringhelli F, Apetoh L, Tesniere A, Janeway Jr CA, Travers P, Walport M, Shlomchik MJ 2001)

Ways to Strengthen Your Immune System

- Don't smoke.

- Eat a diet high in fruits and vegetables.

- Exercise regularly.

- Maintain a healthy weight.

- If you drink alcohol, drink only in moderation.

- Get adequate sleep.

- Take steps to avoid infection, such as washing your hands frequently and cooking meats thoroughly.

- Try to minimize stress.

- Keep current with all recommended vaccines. Vaccines prime your immune system to fight off infections before they take hold in your body.

Many products on store shelves claim to boost or support immunity. But the concept of boosting immunity actually makes little sense scientifically. In fact, boosting the number of cells in your body — immune cells or others — is not necessarily a good thing. For example, athletes who engage in "blood doping" — pumping blood into their systems to boost their number of blood cells and enhance their performance — run the risk of strokes. Attempting to boost the cells of your immune system is especially complicated because there are so many different kinds of cells in the immune system that respond to so many different microbes in so many ways. Which cells should you boost, and to what number? So far, scientists do not know the answer. What is known is that the body is continually generating immune cells.

Certainly, it produces many more lymphocytes than it can possibly use. The extra cells remove themselves through a natural process of cell death called apoptosis— some before they see any action, some after the battle is won. No one knows how many cells or what the best mix of cells the immune system needs to function at its optimum level. (Harvard Health 2021)

So, let's be clear science is undecided on this whole notion of boosting or strengthening your immune system; however, numerous sources suggest resting, eating healthy, exercising, not smoking, etc., as ways to help your immune system. In regard to herbs which many herbalists state can help your immune system create an immune response to deadly pathogens and or create this notion of a preventive measure, are not being honest. "Immunity is a much-abused word that people do not fully understand. The immune system is very complex. These claims about boosting immunity are irrational and unscientific," said Ram Vishwakarma, a noted immunologist and former director of the Council of Scientific & Industrial Research's Indian Institute of Integrative Medicine." (Ramesh & Basu 2020)

"Walk into a store, and you will find bottles of pills and herbal preparations that claim to "support immunity" or otherwise boost the health of your immune system. Although some preparations have been found to alter some components of immune function, thus far, there is no evidence that they actually bolster immunity to the point where you are better protected against infection and disease. Demonstrating whether an herb — or any substance, for that matter — can enhance immunity is, yet a highly complicated matter. Scientists don't know, for example, whether an herb that seems to raise the levels of antibodies in the blood is actually doing anything beneficial for overall immunity." (Harvard Health 2021) Herbs can aid in strengthening the immune system but not prevent a virus from causing a disease the more we know the better informed we will be.

As far as diet specifically, there is no scientific data that will conclude that we can boost our immune systems through eating healthy. This is simply a myth that has been commercialized and passed around social media sites as if it's a reality. Eating has its benefits, but we have absolutely no idea of what foods to eat to prevent a particular virus from infecting us or causing disease. If so, that would mean our food would have to be evolutionary-minded and preventative and smart enough to properly help the immune system recognize millions of viruses and potentially novel viruses we have yet to encounter. That would be a tough feat and has been impossible for hundreds of thousands of years.

With the emergence of Sars-CoV2 and the influence of one of my teachers stressing that for people to properly understand how vaccines work and how viruses behave one needs to know more about the Immune System. This is what I learned before the pandemic; the immune system must be put into proper context because it is our greatest defense against a foreign pathogen. Failure to understand how the immune system functions can lead to some very serious health issues. What Sars-Cov2 taught me about the immune system was simple no matter how healthy you eat or how much you work out you cannot escape infection, and everyone is susceptible of being infected. Herbs can aid in strengthening the immune system but not prevent a virus from causing a disease the more we know the better informed we will be.

Chapter II

Review

1). What is the Immune System?

2). Can you name the 3 different types of Immunity?

3). People whose immune system that has limited ability to fight off infection or disease would be considered as _____.

4). True or False: The adaptive immune system creates an army of immune cells specifically designed to attack that antigen.

5). Can boogers help block pathogens from entering through your nose?

6). True or False: Infants have passive immunity because they are born with antibodies that are transferred through the placenta from their mother.

7). True or False: Adaptive immunity also includes a "memory" that makes future responses against a specific antigen more efficient.

8). Which white blood cell dissolve invaders?

9). True or False: One type of phagocyte is the neutrophil, which fights bacteria.

10). Name 3 ways you can strengthen your immune system.

11). Yes, or No: Is there a difference between self and non-self? Explain your answer.

12). Can herbs or food cause an immune response against a novel virus?

13). True or False: Immunity is a much abused word that people do not fully understand.

Chapter II Sources

- "Immune System (for Parents) - Nemours KidsHealth." Edited by Larissa Hirsch, KidsHealth, The Nemours Foundation, Oct. 2019, kidshealth.org/en/parents/immune.html.

- Davis, Charles P. "Medical Definition of Immune System." MedicineNet, MedicineNet, 29 Mar. 2021, www.medicinenet.com/immune_system/definition.htm.

- Health, National Institute. "NCI Dictionary of Cancer Terms." National Cancer Institute, 2021, www.cancer.gov/publications/dictionaries/cancer-terms/def/immune-system.

- Newman, Tim. "The Immune System: Cells, Tissues, Function, and Disease." Medical News Today, MediLexicon International, 11 Jan. 2018, www.medicalnewstoday.com/articles/320101#In-a-nutshell.

- Medicine, J H. "The Immune System." Johns Hopkins Medicine, John Hopkins Medicine, 2021, www.hopkinsmedicine.org/health/conditions-and-diseases/the-immune-system.

- Institue, National Health. "Overview of the Immune System." National Institute of Allergy and Infectious Diseases, U.S. Department of Health and Human Services, 2021, www.niaid.nih.gov/research/immune-system-overview.

- Institute, National Health. "NCI Dictionary of Cancer Terms." National Cancer Institute, 2021, www.cancer.gov/

- publications/dictionaries/cancer-terms/def/immunocompromised.

- Institue, National Health. "Immune Response: MedlinePlus Medical Encyclopedia." MedlinePlus, U.S. National Library of Medicine, 25 Mar. 2021, medlineplus.gov/ency/article/000821.htm#:~:text=Innate%2C%20or%20nonspecific%2C%20immunity%20is,materials%20from%20entering%20your%20body.

- Arizona , University of. Introduction to Immunology Tutorial, 24 Mar. 2000, www.biology.arizona.edu/immunology/tutorials/immunology/page3.html#:~:text=Adaptive%20immunity%20refers%20to%20antigen,designed%20to%20attack%20that%20antigen.

- Warrington R, Watson W, Kim HL, Antonetti FR. An introduction to immunology and immunopathology. Allergy Asthma Clin Immunol. 2011. doi:10.1186/1710-1492-7-S1-S1.

- Van Parijs L, Abbas AK. Homeostasis and self-tolerance in the immune system: turning lymphocytes off. Science. 1998;280(5361):243-248.

- Sakaguchi S. Naturally arising CD4+ regulatory T cells for immunologic self-tolerance and negative control of immune responses. Rev Immunol. 2004;22:531-562.

- Nikolich-Žugich J. Ageing and life-long maintenance of T-cell subsets in the face of latent persistent infections. Nat Rev Immunol. 2008;8(7):512-522.

- Health, Harvard. "How to Boost Your Immune System." *Harvard Health*, Harvard Health Publishing, 15 Feb. 2021, www.health.harvard.edu/staying-healthy/how-to-boost-your-immune-system.

- Ramesh, Sandhya, et al. "Immunity Boosters Are a Myth - Why You Shouldn't Believe Claims That Promise to Fight Covid." *ThePrint*, 4 Aug. 2020, theprint.in/health/immunity-boosters-are-a-myth-why-you-shouldnt-believe-claims-that-promise-to-fight-covid/470202/.

Covid19 (Symptons/Spread), Transmission, and More
Chapter III

Before December 2019, the average person's understanding of what Covid-19 was and how it spreads was limited. Scientists who study diseases had no clue initially and worked around the clock for potential treatments to prevent the spread of the disease from altering the outcomes of lives that could be lost. From its earliest onsite, Doctors and Hospitals began to learn slowly about how the disease had a wide range of symptoms. As went by, the list began to grow, and the more we would learn in the coming days, weeks, months, and years. "A novel coronavirus, designated as 2019-nCoV, emerged in Wuhan, China, at the end of 2019. As of January 24, 2020, at least 830 cases had been diagnosed in nine countries: China, Thailand, Japan, South Korea, Singapore, Vietnam, Taiwan, Nepal, and the United States. Twenty-six fatalities occurred, mainly in patients who had serious underlying illness. Although many details of the emergence of this virus — such as its origin and its ability to spread among humans — remain unknown, an increasing number of cases appear to have resulted from human-to-human transmission. Given the severe acute respiratory syndrome coronavirus (SARS-CoV) outbreak in 2002 and the Middle East respiratory syndrome coronavirus (MERS-CoV) outbreak in 2012, 2019-nCoV is the third coronavirus to emerge in the human population in the past two decades — an emergence that has put global public health institutions on high alert." (Munster, Ph.D. 2020)

We spoke briefly about what Sars-CoV2 (the virus) that causes Coronavirus 19 (the disease) was, but we will focus on the disease and how it spreads in this chapter. It is very important that we understand the disease that has, till this day, affected hundreds of millions of lives across this globe. In March of 2020 WHO (World Health Organization) declared that Covid 19 outbreak was indeed a pandemic. The U.S. around that same time also made that declaration as well, and while scientists knew very little about the virus and the disease the world began to in some ways panic.

So, what is Coronavirus-19? "Coronavirus disease 2019 (COVID-19) is defined as illness caused by a novel coronavirus called severe acute respiratory syndrome coronavirus 2 (SARS-CoV-2; formerly called 2019-nCoV), which was first identified amid an outbreak of respiratory illness cases in Wuhan City, Hubei Province, China. It was initially reported to the World Health Organization (WHO) on December 31, 2019. On January 30, 2020, the WHO declared the COVID-19 outbreak a global health emergency. On March 11, 2020, the WHO declared COVID-19 a global pandemic, its first such designation since declaring H1N1 influenza a pandemic in 2009." (Cennimo 2021) Since the evolution of virology, scientists have been studying all types of animals to prepare the world for the next explosive pandemic. Epidemiologist assured the public years ago that it was only a matter of time before another potential threat would emerge.

However, "Coronaviruses are important human and animal pathogens." (Hirsch 2021) Because we need to know how this virus behaves in animals just as well as in our bodies. For the rest of this chapter, we will follow the science and concentrate on the cause and effects of Covid-19 and more. "Coronaviruses are important human and animal pathogens. During epidemics, common cold coronaviruses are the cause of up to one-third of community-acquired upper respiratory tract infections in adults and probably also play a role in severe respiratory infections in both children and adults. In addition, certain common cold coronaviruses may cause diarrhea in infants and children. Their role in central nervous system diseases, except for a single case report of encephalitis in a severely immunocompromised infant, has been suggested but not proven." (Hirsch 2021) As of today, Covid-19 cases have slowed down but are very close to 180 million people worldwide. The majority of positive Covid-19 cases are reported as asymptomatic (mild symptoms), and the recovery rate of those diagnosed with Covid-19 is around 160 million people. We will later discuss exactly what those symptoms are.

"Since the first reports of cases from Wuhan, a city in the Hubei Province of China, at the end of 2019, cases have been reported in all continents. The reported case counts underestimate the overall burden of COVID-19, as only a fraction of acute infections is diagnosed and reported. Seroprevalence surveys in the United States and Europe have suggested that after accounting for potential false positives or negatives, the rate of prior exposure to SARS-CoV-2, as reflected by seropositivity, exceeds the incidence of reported cases by approximately 10-fold or more." (Hirsch 2021) In a recent study, it had been reported that the number of deaths is a lot higher than what the world has estimated. Globally we are approaching that 4 million marks and to think it is honestly higher is very disturbing.

Symptoms

Signs and symptoms of coronavirus disease 2019 (COVID-19) may appear 14 days after exposure. This time after exposure and before having symptoms is called the incubation period. Common signs and symptoms can include:

- Fever
- Cough
- Tiredness

Early symptoms of COVID-19 may include a loss of taste or smell.

Other symptoms can include:

- Shortness of breath or difficulty breathing
- Muscle aches
- Chills
- Sore throat
- Runny nose
- Headache
- Chest pain
- Pink eye (conjunctivitis)
- Nausea
- Vomiting
- Diarrhea
- Rash

This list is not all-inclusive. Children have similar symptoms to adults and generally have mild illness. The severity of COVID-19 symptoms can range from very mild to severe. Some people may have only a few symptoms, and some people may have no symptoms at all. Some people may experience worsened symptoms, such as worsened shortness of breath and pneumonia, about a week after symptoms start.

People who are older have a higher risk of serious illness from COVID-19, and the risk increases with age. People who have existing medical conditions also may have a higher risk of serious illness. Certain medical conditions that may increase the risk of serious illness from COVID-19 include:

- Serious heart diseases, such as heart failure, coronary artery disease, or cardiomyopathy

- Cancer

- Chronic obstructive pulmonary disease (COPD)

- Type 1 or type 2 diabetes

- Overweight, obesity or severe obesity

- High blood pressure

- Smoking

- Chronic kidney disease

- Sickle cell disease or thalassemia

- Weakened immune system from solid organ transplants

- Pregnancy

- Asthma

- Chronic lung diseases such as cystic fibrosis or pulmonary fibrosis

- Liver disease

- Dementia

- Down syndrome

- Weakened immune system from bone marrow transplant, HIV, or some medications

- Brain and nervous system conditions

• Substance use disorders

This list is not all inclusive. Other underlying medical conditions may increase your risk of serious illness from COVID-19. List accredited to the Mayo Clinic (1998-2021 Mayo Foundation for Medical Education and Research (MFMER). All rights reserved.)

How It Spreads

COVID-19 is spread in three main ways:

- Breathing in the air when close to an infected person who is exhaling small droplets and particles that contain the virus.

- Having these tiny droplets and particles that contain virus land on the eyes, nose, or mouth, especially through splashes and sprays like a cough or sneeze.

- Touching eyes, nose, or mouth with hands that have the virus on them. It is essential to wash your hands before touching your mouth, nose, face, or eyes.

- People can spread the COVID-19 disease to each other.

- Infected people may be able to spread the disease before they have symptoms or feel sick.

- A person can also spread the disease if they have no symptoms. Research has shown that around 40-50% of people infected do not develop symptoms.

(Minnesota Department of Health 2021)

The transmission of Covid19 has been extensively studied, and over the past year, the science has done a great job of fully understanding how transmission occurs. "Viral transmission is the process by which viruses spread between hosts. It includes spread to members of the same host species or spread to different species in the case of viruses that can cross species barriers." (Journal of Nature 2021)

"Route of person-to-person transmission — Direct person-to-person respiratory transmission is the primary means of transmission of severe acute respiratory syndrome coronavirus 2 (SARS-CoV-2). It is thought to occur mainly through close-range contact (i.e., within approximately six feet or two meters) via respiratory particles; virus released in the respiratory secretions when a person with infection coughs, sneezes, or talks can infect another person if it is inhaled or makes direct contact with the mucous membranes. Infection might also occur if a person's hands are contaminated by these secretions or by touching contaminated surfaces, and then they touch their eyes, nose, or mouth, although contaminated surfaces are not thought to be a major route of transmission. SARS-CoV-2 can also be transmitted long distances through the airborne route (through inhalation of particles that remain in the air over time and distance), but the extent to which this mode of transmission has contributed to the pandemic is uncertain. Scattered reports of SARS-CoV-2 outbreaks (e.g., in a restaurant, on a bus) have highlighted the potential for longer-distance airborne transmission in enclosed, poorly ventilated spaces. Experimental studies have also supported the feasibility of airborne transmission. As examples, studies using specialized imaging to visualize respiratory exhalations have suggested that respiratory droplets may get aerosolized or carried in a gas cloud and have horizontal trajectories beyond six feet (two meters) with speaking, coughing, or sneezing. Other studies have identified viral RNA in ventilation systems and in air samples of hospital rooms of patients with COVID-19, including patients with mild infection; attempts to find viable virus in air and surface specimens in health care settings have only rarely been successful.

Nevertheless, the overall transmission and secondary attack rates of SARS-CoV-2 suggest that long-range airborne transmission is not a primary mode. Furthermore, in a few reports of health care workers exposed to patients with undiagnosed infection while using only contact and droplet precautions, no secondary infections were identified despite the absence of airborne precautions. Reflecting the current uncertainty regarding the relative contribution of different transmission mechanisms, recommendations on airborne precautions in the health care setting vary by location; airborne precautions are universally recommended when aerosol-generating procedures are performed." (McIntosh MD 2021)

"The emergence and spread of a novel coronavirus (2019-nCoV) from Wuhan, China, has become a global health concern. Since the detection of the coronavirus in late December 2019, several countries have reported sporadic imported cases among travelers returning from China. We report one family cluster of 2019-nCoV originating from a Chinese man. On January 22, 2020, a 65-year-old man with a history of hypertension, type 2 diabetes, coronary heart disease for which a stent had been implanted, and lung cancer was admitted to the emergency department of Cho Ray Hospital, the referral hospital in Ho Chi Minh City, for low-grade fever and fatigue. He had become ill with fever on January 17, a total of 4 days after he and his wife had flown to Hanoi from the Wuchang district in Wuhan, where outbreaks of 2019-nCoV were occurring. He reported that he had not been exposed to a "wet market" (a market where dead and live animals are sold) in Wuhan. Throat swabs obtained from the patient tested positive for 2019-nCoV on real-time reverse-transcription–polymerase-chain-reaction (RT-PCR) assays. On admission to the hospital, the man was isolated and treated empirically with antiviral agents, broad spectrum antibiotics, and supportive therapies. Chest radiographs obtained on admission showed an infiltrate in the upper lobe of the left lung. On January 25, he received supplemental oxygen through a nasal cannula at a rate of 5 liters per minute because of increasing dyspnea with hypoxemia. The partial pressure of oxygen was 57.2 mm Hg while he was breathing ambient air, and a progressive infiltrate, and consolidation was observed on chest radiographs. His fever disappeared on January 25, and his clinical condition has improved since January 26. His wife had no symptoms of illness while they were traveling. She was healthy as of January 28. (Phan, Ph.D. 2020)

Transmission is inevitable and if we are going to expose ourselves to certain environmental conditions, we must fully understand how the virus maneuvers. Based on the available data above regarding person-to-person transmission, whether it be airborne transmission or otherwise. Viral Shedding is another form of transmission that is why scientists have encouraged quarantining up to 14 days during the initial onset, but after further studies suggested a minimum of 10 days. "When an individual gets infected by a respiratory virus like SARS-CoV-2, the virus particles will bind to the various types of viral receptors, particularly the angiotensin-converting enzyme 2 (ACE2) receptors in the case of SARS-CoV-2, that line the respiratory tract. Throughout this ongoing process, infected individuals, who may not yet be experiencing any of the viral symptoms, are shedding viral particles while they talk, exhale, eat, and perform other normal daily activities. Under normal circumstances, viral shedding will not persist for more than a few weeks; however, as researchers gain a more in-depth understanding of the viral clearance of SARS-CoV-2, they have found that certain populations will shed this virus for much longer durations." (Cuffari 2021)

"Viral shedding and period of infectiousness — The precise interval during which an individual with SARS-CoV-2 infection can transmit infection to others is uncertain. The potential to transmit SARS-CoV-2 begins prior to the development of symptoms and is highest early in the course of illness; the risk of transmission decreases thereafter. Transmission after 7 to 10 days of illness is unlikely, particularly for otherwise immunocompetent patients with non-severe infection. Period of greatest infectiousness – Infected individuals are more likely to be contagious in the earlier stages of illness when viral RNA levels from upper respiratory specimens are the highest. One modeling study, in which the mean serial interval between the onset of symptoms among 77 transmission pairs in China was 5.8 days, estimated that infectiousness peaked between two days before and one day after symptom onset and declined within seven days. In another study that evaluated over 2500 close contacts of 100 patients with COVID-19 in Taiwan, all of the 22 secondary cases had their first exposure to the index case within six days of symptom onset; there were no infections documented in the 850 contacts whose exposure was after this interval. Prolonged viral RNA detection does not indicate prolonged infectiousness – The duration of viral RNA shedding is variable and may increase with age and the severity of illness. In a review of 28 studies, the pooled median duration of viral RNA detection in respiratory specimens was 18 days following the onset of symptoms; in some individuals, viral RNA was detected from the respiratory tract several months after the initial infection. Detectable viral RNA, however, does not necessarily indicate the presence of infectious virus, and there appears to be a threshold of viral RNA level below which infectiousness is unlikely. For example, in a study of nine patients with mild COVID-19, infectious virus was not detected from respiratory specimens when the viral RNA level was <106 copies/mL. In other studies, infectious virus has only been detected in respiratory specimens with high concentrations of viral RNA. Such high viral RNA concentrations are reflected by lower numbers of reverse transcriptase-polymerase chain reaction (RTPCR) amplification cycles necessary for detection. Depending on the study, the cycle threshold (Ct) for specimen culture positivity may vary from <24 to ≤32. According to information from the United States Centers for Disease Control and Prevention (CDC), by three days after clinical recovery, if viral RNA is still detectable in upper respiratory specimens. The RNA concentrations are generally at or below the levels at which replication-competent virus can be reliably isolated; additionally, isolation of infectious virus from upper respiratory specimens more than 10 days after illness onset has only rarely been documented in patients who had non-severe infection and whose symptoms have resolved. Except for sporadic reports of reinfection, infectious virus has not been isolated from respiratory specimens of immunocompetent patients who have a repeat positive RNA test following clinical improvement and initial viral RNA clearance, and transmission from such patients has not been documented. (McIntosh MD 2021)

In my opinion, Dr. McIntosh does a great job of outlining the scientific data that exists regarding transmission and how the test revealed certain aspects of each level of viral RNA and how it is detected. He continued to follow the science and not his opinion even when he refers to reinfection. This is important because scientific data is limited regarding reinfections. However, he continues, "Risk of transmission depends on exposure type — The risk of transmission from an individual with SARS-CoV-2 infection varies by the type and duration of exposure, use of preventive measures, and likely individual factors (eg, the amount of virus in respiratory secretions). Many individuals do not transmit SARS-CoV-2 to anyone else, and epidemiological data suggest that the minority of index cases result in the majority of secondary infections. The risk of transmission after contact with an individual with COVID-19 increases with the closeness and duration of contact and appears highest with prolonged contact in indoor settings." (McIntosh MD 2021) This is important because most people assume that they aren't susceptible to even contracting the virus and if they actually do, they will presumably have little to no symptoms.

Earlier, when we discussed the number of people who have actually recovered from Covid19, we often hear a few words called asymptomatic and symptomatic. Most people assume they are asymptomatic and can fight off infection naturally because they develop mild or little complications. However, if we do not properly define those terms, we could get lost in their true meaning.

Defining asymptomatic and pre-symptomatic: "Asymptomatic is someone who has the infection but no symptoms and will not develop them later. Pre-symptomatic is someone who has the infection but don't have any symptoms yet. Both groups can spread the infection. COVID-19 spreads easily, and we believe that's because it's spread by those who don't know they're infected. We suspect that individuals who are pre-symptomatic are infectious for two to three days before having symptoms." (Meller 2021)

"The biologic basis for this is supported by a study of a SARS-CoV-2 outbreak in a long-term care facility, in which infectious virus was cultured from RT-PCR-positive upper respiratory tract specimens in pre-symptomatic and asymptomatic patients as early as six days prior to the development of typical symptoms. The levels and duration of viral RNA in the upper respiratory tract of asymptomatic patients are also similar to those of symptomatic patients. The risk of transmission from an individual who is asymptomatic appears less than that from one who is symptomatic. As an example, in an analysis of 628 COVID-19 cases and 3790 close contacts in Singapore, the risk of secondary infection was 3.85 times higher among contacts of a symptomatic individual compared with contacts of an asymptomatic individual. Similarly, in an analysis of American passengers on a cruise ship that experienced a large SARS-CoV-2 outbreak, SARS-CoV-2 infection was diagnosed in 63 percent of those who shared a cabin with an individual with asymptomatic infection, compared with 81 percent of those who shared a cabin with a symptomatic individual and 18 percent of those without a cabin mate. Nevertheless, asymptomatic or pre-symptomatic individuals are less likely to isolate themselves from other people, and the extent to which transmission from such individuals contributes to the pandemic is uncertain. A CDC modeling study estimated that 59 percent of transmission could be attributed to individuals without symptoms: 35 percent from pre-symptomatic individuals, and 24 percent from those who remained asymptomatic. This estimate was based on several assumptions, including that 30 percent of infected individuals never develop symptoms and are 75 percent as infectious as those who do. (McIntosh 2021)

What I am explaining here are the things that Sars-Cov2 taught me which are: How to properly define certain words that I was unfamiliar with before but make them more inclusive in my learning while analyzing data and understanding how scientists come to conclusions based on information collected summarized and explained using the scientific method. From the onset, people have dismissed asymptomatic as someone who has gotten infected and had no symptoms at all, but most people do not consider the lives around them just because they did not have a certain reaction or illness. Your mother, father, grandmother, grandfather, aunt, uncle, sister, brother, cousin, friend, neighbor, etc. etc.… May be severely infected by your ability to transmit the virus from person to person, and now you have weaponized Covid19 and potentially risk the lives of someone you hold near and dear to your heart based on your own ignorance.

In a study published by BioFrontiers Institute at the University of Colorado, they concluded that 2% of the population was responsible for 90% of virus that circulates. "We analyze data from the fall 2020 pandemic response efforts at the University of Colorado Boulder, where more than 72,500 saliva samples were tested for severe acute respiratory syndrome coronavirus 2 (SARS-CoV-2) using qRT-PCR. All samples were collected from individuals who reported no symptoms associated with COVID-19 on the day of collection. From these, 1,405 positive cases were identified. The distribution of viral loads within these asymptomatic individuals was indistinguishable from what has been previously observed in symptomatic individuals. Regardless of symptomatic status, ~50% of individuals who test positive for SARS-CoV-2 seem to be in noninfectious phases of the disease, based on having low viral loads in a range from which live virus has rarely been isolated. We find that, at any given time, just 2% of individuals carry 90% of the virions circulating within communities, serving as viral "super-carriers" and possibly also super-spreaders." (Yang, Saldi, Gonzales 2021).

This study supports the notion that asymptomatic carries of Sars-CoV2, especially new variants, are weaponizing a virus that can end the lives of those of us who have taken precautionary measures who take the potential threat of this novel virus serious.

Studying the disease

CDC and other agencies and institutions around the world are conducting thousands of epidemiological studies to learn more about COVID-19 and the virus that causes it. These studies help us understand:

• The time between when someone is exposed to the virus and when they have symptoms (incubation period). We now know that someone can be infected with the virus for 2–14 days before they feel sick and that some people never feel sick.

• How long a person who is infected can shed (release from the body) the virus. To avoid spreading infection, we recommend that people infected with the virus avoid being around others until they have gone 3 days without fever, their symptoms have cleared, and 10 days have passed since their symptoms started.

• The range of signs, symptoms, and severity of the disease (spectrum of disease). Knowing this information helps people be on the lookout for early symptoms and helps healthcare professionals diagnose and treat the disease.

• The risk factors associated with severe disease. We now know that older people who have serious chronic health conditions are at higher risk of becoming very sick from COVID-19.

• How often the disease causes illness and death in a population (morbidity and mortality rate). This information helps epidemiologists understand the impact of COVID-19 on public health.

(National Center for Immunization and Respiratory Diseases

(NCIRD), Division of Viral Diseases 2021)

Antibody (Serology) Testing for COVID-19: Information for Patients and Consumers "SARS-CoV-2 antibody (often referred to as serology) tests look for antibodies in a sample to determine if an individual has had a past infection with the virus that causes COVID-19. COVID-19 antibody tests can help identify people who may have been infected with the SARS-CoV-2 virus or have recovered from a COVID-19 infection. At this time, researchers do not know whether the presence of antibodies means that you are immune to COVID-19; or if you are immune, how long it will last. In people who have received a COVID-19 vaccination, antibody testing is not recommended to determine whether you are immune or protected from COVID-19. Many antibody tests are currently in development or available for use to detect antibodies to SARS-CoV-2. However, not all antibody tests that are being marketed to the public have been evaluated and authorized by the FDA." (FDA 2021)

What tests are used to diagnose COVID-19?

The FDA approved these types of tests for diagnosing a COVID-19 infection:

PCR test. Also called a molecular test, this COVID-19 test detects genetic material of the virus using a lab technique called polymerase chain reaction (PCR). A fluid sample is collected with a nasal swab or a throat swab, or you may spit into a tube to produce a saliva sample. Results may be available in minutes if analyzed onsite or a few days — or longer in locations with test processing delays — if sent to an outside lab. PCR tests are very accurate when properly performed by a health care professional, but the rapid test can miss some cases.

Antigen test. This COVID-19 test detects certain proteins in the virus. Using a nasal swab to get a fluid sample, antigen tests can produce results in minutes. Others may be sent to a lab for analysis. A positive antigen test result is considered accurate when instructions are carefully followed, but there's an increased chance of false-negative results — meaning it's possible to be infected with the virus but have a negative result. Depending on the situation, the doctor may recommend a PCR test to confirm a negative antigen test result.

The FDA granted emergency use authorization for certain home COVID-19 test kits. You can collect your own sample of nasal fluid or saliva at home. Most of these tests require a doctor's prescription. The accuracy of these tests varies, so a negative test does not completely rule out having the COVID-19 virus. Only get an at-home test that's authorized by the FDA or approved by your local health department. (Mayo Clinic 2021)

However, "People who have been infected with a virus might develop antibodies (which are proteins in the blood that fight the virus) even if they don't know they are infected. Serologic tests can be used to detect antibodies. By counting the number of people with antibodies to COVID-19, scientists can learn how much the disease has spread in a population. Antibody tests are useful because they include infections that might have been missed because people had no symptoms (were asymptomatic) or mild symptoms and therefore did not get tested or receive medical care. Antibody tests help answer other important questions about how COVID-19 infections are progressing through populations over time and help estimate how much of the population has not yet been infected, helping public health officials plan for healthcare needs." (CDC 2021)

Covid19 taught me to pay attention to the growing amount of data that exist each day. It also made me slow down and review that day on a case-by-case basis, understand scientific terminology, and expect the science to change as more information became readily available based on scientific prioritization. Understanding how viruses transmit was key because some people assume it's just by contact, not considering all the other factors that exist when it comes to the spread of viruses. Also, understanding the difference between asymptomatic, pre-symptomatic, and symptomatic regarding a novel viruses shows just how easily this virus impacted the world so suddenly. Most importantly, people assumed that Covid19 was just like all other viruses. The layman has never considered how these viruses behave and attack the overall body; it is more than just the respiratory system that is at the mercy of the virus, and this is What Sars-Cov2 taught me!

Chapter III
Review

1). When did WHO officially declare the outbreak of Covid19 a pandemic?

2). Can you name 4 symptoms caused by Covid19?

3). Name 3 ways Covid19 spreads:

4). _____ is the process by which viruses spread between hosts. It includes spread to members of the same host species or spread to different species in the case of viruses that can cross species barriers.

5). What is Viral Shedding?

6). True or False: Infected individuals are more likely to be contagious in the earlier stages of illness when viral RNA levels from upper respiratory specimens are the highest.

7). Explain Risk of Infection:

8) Define Asymptomatic:

9). _____ is someone who has the infection but don't have any symptoms yet

10). How long a person who is infected can shed (release from the body) the virus?

11). What tests are used to diagnose COVID-19?

12). What are antibodies?

Chapter III Sources

- Control & Prevention, Center Disease. "Symptoms of COVID-19." Centers for Disease Control and Prevention, Centers for Disease Control and Prevention, 2021, www.cdc.gov/coronavirus/2019-ncov/symptoms-testing/symptoms.html.

- Staff, Mayo Clinic. "Coronavirus Disease 2019 (COVID-19)." Mayo Clinic, Mayo Foundation for Medical Education and Research, 2 June 2021, www.mayoclinic.org/diseases-conditions/coronavirus/symptoms-causes/syc-20479963.

- David J Cennimo, MD. "Coronavirus Disease 2019 (COVID-19)." Practice Essentials, Background, Route of Transmission, Medscape, 20 May 2021, emedicine.medscape.com/article/2500114-overview.

- McIntosh, Kenneth. UpToDate, 9 June 2021, www.uptodate.com/contents/covid-19-epidemiology-virology-and-prevention#H2549483976.

- Caliendo, Angela, and Kimberly Hanson. Edited by Martin Hirsch and Allyson Bloom, UpToDate, UpToDate, 16 Apr. 2021, www.uptodate.com/contents/covid-19-diagnosis?topicRef=8298&source=see_link.

- Health, Minn Dept. "About COVID-19." Minnesota Dept. of Health, 8 June 2021, www.health.state.mn.us/diseases/coronavirus/basics.html.

- Division of Viral Diseases, National Center for Immunization and Respiratory Diseases (NCIRD). "Studying the Disease." Centers for Disease Control and Prevention, Centers for Disease Control and Prevention, 1 July 2020, www.cdc.gov/coronavirus/2019-ncov/science/about-epidemiology/

- studying-the-disease.html?CDC_AA_refVal=https%3A%2F%2Fwww.cdc.gov%2Fcoronavirus%2F2019-ncov%2Fcases-updates%2Fabout-epidemiology%2Fstudying-the-diesease.html.

- Nature, Journal of. Nature News, Nature Publishing Group, 2020, www.nature.com/subjects/viral-transmission.

- Cuffari, Benedette. "What Is Viral Shedding?" News, 16 Mar. 2021, www.news-medical.net/health/What-is-Viral-Shedding.aspx#:~:text=Throughout%20this%20ongoing%20process%2C%20infected,perform%20other%20normal%20daily%20activities.

- Health, National Institute. "Asymptomatic: MedlinePlus Medical Encyclopedia." MedlinePlus, U.S. National Library of Medicine, 2021, medlineplus.gov/ency/article/002217.htm.

- Meller, Megan. "The Asymptomatic and Pre-Symptomatic Spread of COVID-19." Gundersen Health System, 2020, www.gundersenhealth.org/covid19/the-asymptomatic-and-pre-symptomatic-spread-of-covid-19/.

- and Radiological Health, Center for Devices. "Antibody (Serology) Testing for COVID-19." U.S. Food and Drug Administration, FDA, 19 May 2021, www.fda.gov/medical-devices/coronavirus-covid-19-and-medical-devices/antibody-serology-testing-covid-19-information-patients-and-consumers.

- Marshall III, William M. "Here's What You Need to Know about COVID-19 Testing." Mayo Clinic, Mayo Foundation for Medical Education and Research, 12 Dec. 2020, www.mayoclinic.org/diseases-conditions/coronavirus/expert-answers/covid-antibody-tests/faq-20484429.

- Munster, Vincent J., et al. "A Novel Coronavirus Emerging in China - Key Questions for Impact Assessment: NEJM." New England Journal of Medicine, 21 Apr. 2021, www.nejm.org/doi/full/10.1056/NEJMp2000929?query=featured_coronavirus.

- Shimabukuro and Others, T.T., et al. "Importation and Human-to-Human Transmission of a Novel Coronavirus in Vietnam: NEJM." New England Journal of Medicine, 21 Apr. 2021, www.nejm.org/doi/full/10.1056/NEJMc2001272?query=featured_coronavirus.

- Yang, Qing, et al. Just 2% of SARS-CoV-2–Positive Individuals Carry 90% of the Virus Circulating in Communities, no. PNAS, 11 Apr. 2021, pp. 1–6., doi: https://doi.org/10.1073/pnas.2104547118.

**The Damage Covid Causes
Chapter IV**

Since the emergence of Sars-CoV2, the virus caused by the disease Covid-19 we knew very little about the long-term effects of the virus. We knew in name alone that this virus was a respiratory virus that was given, but the way Covid ravaged the body took Doctors by surprised and off guard. This would lead to a series of experimental behaviors looking to contain and control the circumstances and prevent the death of numerous lives. I remember reading articles where steroids were used, a series of preventatives for other treatments, and more. But to no avail, lives continued to be lost! This novel virus, from its onset, showed that nature was in full control despite the scientific efforts made by the world. Months would pass before science would catch up to Covid-19.

"Most people who have coronavirus disease 2019 (COVID-19) recover completely within a few weeks. But some people — even those who had mild versions of the disease — continue to experience symptoms after their initial recovery. These people sometimes describe themselves as "long haulers," and the conditions have been called post-COVID-19 syndrome or "long COVID-19." These health issues are sometimes called post-COVID-19 conditions. They're generally considered to be effects of COVID-19 that persist for more than four weeks after you've been diagnosed with the COVID-19 virus. Older people and people with many serious medical conditions are the most likely to experience lingering COVID-19 symptoms, but even young, otherwise healthy people can feel unwell for weeks to months after infection." (Mayo Clinic 2021)

The aftermath of numerous months of increased death didn't give scientists nor doctors time to focus on what stared them right in their faces. The long-term effects of Covid-19 left many who recovered with new health issues that were unknown to them prior to naturally being infected by Sars-CoV2. In this chapter, we will review the numerous health issues people have endured but not limited to hospitalization after recovering from Covid-19.

"Researchers are growing miniature organs in the laboratory to study how the new coronavirus ravages the body. Studies in these organoids are revealing the virus's versatility at invading organs, from the lungs to the liver, kidneys, and gut." (Mallapaty 2020) This study was published in June of 2020, literally 6 months after the first known U.S. case was acknowledged. Around that same time, the U.S. infections numbers hit 2 million earlier in the month, with over 100,000 deaths reported. Also, this research is critical for people with pre-existing conditions because it puts them directly in the line of fire with how Covid-19 attacks the body.

Before this study, we knew in February that Covid-19 attacked the lungs but, even more specifically, how fluid would build up in the lungs, making it difficult for oxygen to flow in and out of the lungs. "Two recent studies in the journal Nature provide some of the most detailed analyses yet about the effects on the human body of SARSCoV-2, the coronavirus that causes COVID-19. The research shows that in people with advanced infections, SARS-CoV-2 often unleashes a devastating series of host events in the lungs prior to death. These events include runaway inflammation and rampant tissue destruction that the lungs cannot repair. Both studies were supported by NIH. One comes from a team led by Benjamin Izar, Columbia University, New York. The other involves a group led by Aviv Regev, now at Genentech, and formerly at Broad Institute of MIT and Harvard, Cambridge, MA. Each team analyzed samples of essential tissues gathered from COVID-19 patients shortly after their deaths. Izar's team set up a rapid autopsy program to collect and freeze samples within hours of death. He and his team performed single-cell RNA sequencing on about 116,000 cells from the lung tissue of 19 men and women. Similarly, Regev's team developed an autopsy biobank that included 420 total samples from 11 organ systems, which were used to generate multiple single-cell atlases of tissues from the lung, kidney, liver, and heart. Izar's team found that the lungs of people who died of COVID-19 were filled with immune cells called macrophages. While macrophages normally help to fight an infectious virus, they seemed, in this case, to produce a vicious cycle of severe inflammation that further damaged lung tissue. The researchers also discovered that the macrophages produced high levels of IL-1β, a type of small inflammatory protein called a cytokine. This suggests that drugs to reduce the effects of IL-1β might have promise to control lung inflammation in the sickest patients. As a person clears and recovers from a typical respiratory infection, such as the flu, the lung repairs the damage. But in severe COVID-19, both studies suggest this isn't always possible. Not only does SARS-CoV-2 destroy cells within air sacs, called alveoli, that are essential for the exchange of oxygen and carbon dioxide, but the unchecked inflammation apparently also impairs remaining cells from repairing the damage. In fact, the lungs' regenerative cells are suspended in a kind of reparative limbo, unable to complete the last steps needed to replace healthy alveolar tissue." (Collins M.D. 2021) Both studies presented evidence of long-term complications surrounding patients but also provided a unique understanding moving forward as doctors looked for potential solutions to a problem caused by a novel virus.

"The majority of patients with SARS-CoV-2 infection remain asymptomatic or have mild symptoms, including fever, cough, anosmia, and headache. However, around 15% develop severe pulmonary disease typically over 10 days, leading to respiratory compromise, which might progress to multi-organ failure, coagulopathy, and death. Oxygen supplementation, invasive ventilation, and other supportive measures now form part of the standard of care in hospitalized patients; however, mortality remains high among those with critical disease. Common risk factors consistently associated with severe COVID-19 are now well established and include advancing age, male sex and a burden of comorbidity, including hypertension, heart disease, diabetes, and malignancy." (Marjot, Webb, Barritt, Moon & more 2021)

This is 'What Sars-Cov2 Taught Me' that not only did Covid-19 attack the organs, but it also created future health risk and I paid close attention to how our bodies would behave after being infected with this virus. Once we began to start learning more about Covid-19 causes and effects, I knew that people could home in on solutions to the problem. However, one of the biggest health risks in the black community is hypertension. Many people I know have high blood pressure and are currently on medication, and I needed to know how that would affect people close to me directly.

"COVID-19, the disease caused by the SARS-CoV-2 coronavirus, can damage the heart muscle and affect heart function. There are several reasons for this. The cells in the heart have angiotensin-converting enzyme-2 (ACE-2) receptors where the coronavirus attaches before entering cells. Heart damage can also be due to high levels of inflammation circulating in the body. As the body's immune system fights off the virus, the inflammatory process can damage some healthy tissues, including the heart. Coronavirus infection also affects the inner surfaces of veins and arteries, which can cause blood vessel inflammation, damage to very small vessels, and blood clots, all of which can compromise blood flow to the heart or other parts of the body. "Severe COVID-19 is a disease that affects endothelial cells, which form the lining of the blood vessels," Post says." (Post M.D. M.S. 2020)

This information was critical in my eyes because it would further shed light on how damaging this would be to black and brown people who live in impoverished areas and would likely be susceptible to being infected with Sars-Cov-2. Dr. Post also mentioned that there were still some unanswered questions surrounding people who would potentially develop long-term heart issues. She mentioned that the NIH - National Institute of Health was currently working with dozens of institutions attempting to track patients who have recovered from Covid-19.

Sars-Cov-2 damage did not just stop at our lungs, kidneys, heart, and liver it traveled within the brain, causing what we know now as 'brain fog', which affects your ability to think. This is why some people have a hard time remembering certain things; however, it also affected peoples taste and smell which are leading symptoms recognized by all physicians today. "In a review of case reports from 901 COVID-19 patients, Mark Ellul, at the University of Liverpool, and colleagues reported a range of neurological manifestations, including loss of smell and taste, confusion, encephalitis (inflammation in the brain), and Guillain-Barré syndrome (a disorder in which the immune system attacks the body's nerves) (The Lancet Neurology, published online, 2020). A case report of 58 patients from France described neurological findings in 67% of patients (Helms, J. et al., The New England Journal of Medicine, Vol. 382, No. 23, 2020). The prevalence of neurological problems remains an open question, but it's safe to conclude that "neurological problems are not rare for COVID-19 patients," says Majid Fotuhi, MD, Ph.D., medical director of NeuroGrow Brain Fitness Center and lead author of a comprehensive review of COVID-19's effects on the nervous system (Journal of Alzheimer's Disease, Vol. 76, No. 1, 2020). "Our best estimate so far is that 30% to 50% of hospitalized patients have neurological issues," he says. In their review, Fotuhi and his colleagues describe the variety of neurological complications in patients with COVID-19.

"There's a wide range of symptoms, including headaches, dizziness, weakness, confusion, eye movement problems, seizures, and paralysis," he says. "The two most common neurological problems seem to be stroke and delirium." In general, people who experience more serious symptoms of COVID-19 tend to have more brain-related complications, Fotuhi says. "Broadly speaking, the sicker they are, the more neurological issues they have." But there are exceptions to that rule. A study by scientists in England of 43 patients with severe neurological complications from COVID-19 found that some patients had relatively mild respiratory symptoms (Paterson, R.W., et al., Brain, published online, 2020)." (Weir 2020) Neurologists are attempting to answer more questions surrounding the effects of Sars-Cov-2 on the brain just as other Doctors are in their respective fields.

For people that know me know that I have a sister who previously had a kidney transplant however she has to have a 2nd transplant due to newly discovered failures. I needed to know the severity of the disease regrading transplants because I wanted to make sure my sister would be ok. I began to read a study published in AJMC, and it talked about how deadly viral RNA was found in the kidney of someone who had died from Covid-19. "Authors from 2 German institutions, the University Medical Center Hamburg-Eppendorf and the University of Gottigen, led by Tobias Huber, MD, examined kidney samples from 63 patients who died after developing SARS-CoV-2 respiratory infection. They found SARS-CoV-2 RNA in 60% of the patients, and this RNA presence in the kidneys "was associated with older age and an increased number of coexisting conditions." The research team found that patients with SARS-CoV2 RNA had a shorter span between diagnosis and death, supporting a correlation between the virus attacking organs beyond the respiratory system and death in the first 3 weeks after diagnosis. Among the 63 patients, 39 (62%) already showed kidney problems before death; among this group, 23 of 32 patients with AKI had SARS-CoV-2 RNA present in their kidneys after death. For those without AKI, viral RNA was found in 3 of 7 patients. When the researchers isolated SARSCoV-2 from an autopsied kidney and examined samples in vitro, they were able to produce "a 1000-times increase in viral RNA" in 48 hours, thus showing that the virus remained active even after the patient died. "This suggests that SARS-CoV-2 is able to target the kidney, pointing towards the importance of early urinary testing and eventual therapeutic prevention of kidney infection," they wrote." (Caffrey 2020). This made me a bit nervous because this disease doesn't care about what it encounters. It needs a host and at all costs it can't survive without one and you best believe it will ravage the body unlike any other virus of our time.

"If COVID-19 patients with pneumonia need to be ventilated, this can also damage the kidneys. Acute kidney failure often occurs. Because pneumonia often causes a lot of fluid to accumulate in the lungs, patients are given a drug that removes fluid from the body. However, this reduces the blood supply to the kidneys, and they can no longer fulfill their cleansing function. In addition, the blood coagulates faster in severe COVID-19 disease. As a result, blood clots can easily form, blocking the blood vessels and often also the kidneys. Small infarctions in the kidney tissue have been observed in numerous patients. In about 30% of these patients, the kidneys are acutely restricted to such an extent that they require dialysis. It is not clear yet whether the kidneys heal after the patients recover or whether SARS-Cov-2 triggers long-term damage to the organs." (Freund2020)

Until we know more about if kidneys are actually healing, I'll remain on edge, and an overprotective little brother concerned about his sister. However, I'll remain hopeful that science will answer this question for me as soon as possible.

We battle covid from head to toe literally outside our major organs; you have what many people call covid toes, covid tongue, and covid skin. These are some of the effects of being infected with Sars-Cov2. "Toward the end of March, as San Francisco began to warm up, Sonia got cold feet. She put on wool socks and turned up her heater. Still, her feet felt frozen. Three days later, her soles turned splotchy purple. Red dots appeared on her toes. At night, her cold feet itched and burned. Walking hurt. And she was exhausted, napping through afternoon Zoom meetings. "It was so bizarre," says Sonia, a San Francisco resident. A week later, her symptoms were gone. "Yes, COVID," wrote Lindy Fox, MD, a UCSF professor of dermatology, replying to an email describing Sonia's case. Sonia wasn't surprised. Anyone like her, who's been following news of the pandemic, has probably heard about "COVID toes," a painful or itchy skin rash that sometimes pops up in young adults with otherwise mild or asymptomatic cases of COVID-19. "It looks like what we call pernio, or chilblains," Fox says, "which is a pretty common phenomenon when somebody goes out in cold weather – they start to get purple or pink bumps on their fingers or toes." Many people with rashes like Sonia's don't test positive for COVID-19, Fox says, which has made some clinicians skeptical of the connection; when patients have both, it's just a coincidence, they believe. But Fox doesn't think so. For one thing, "the time of year is wrong," she says. "Pernio usually shows up in the dead of winter." Even more compelling, dermatologists around the world are "getting crazy numbers of calls about it," Fox says. "In the last three weeks, I've had somewhere between 10 and 12 patients. Normally, I have four a year." (Bleicher & Conrad 2020)

Covid toes can develop at any age or sex. You may not pay any attention to the signs at first glance, but they come on quickly and are said to last about 10 to 14 days on average. We still have some unanswered questions regarding covid toes, like is it contagious and will it have any lingering effects on us in the future.

Covid skin can lead to rashes in weird places and in some instances, bumps along the armpits, hands, and leg area. "Many diseases, such as measles and chickenpox, cause a distinctive rash that helps doctors diagnose a patient. COVID-19 is different. There is no single COVID-19 rash. What you may see: You can have COVID-19 and never develop a rash. When a patient with COVID-19 does develop a rash, it can look like any of the following: Patchy rash, Itchy bumps, Blisters that look like chickenpox, round pinpoint spots on the skin, large patch with several smaller ones, A lace-like pattern on the skin, Flat spots and raised bumps that join together." (American Academy of Dermatology 2020) It is also determined that Covid skin lasts about 2 to 12 days in most people, and if it's contagious is unknown.

Covid fingers are like skin rashes; however, some discoloration has occurred in a few instances leading to amputation in an elderly person who developed a more severe case of Covid fingers. "An 86-year-old woman presented in April 2020 with black fingers. She had dry gangrene of the second, fourth, and fifth fingers of the right hand. In March, she had acute coronary syndrome and was put on dual anti-platelet therapy. Because of the COVID-19 pandemic, she underwent oropharyngeal swab testing, which was positive for COVID-19. Arterial duplex ultrasonography showed monophasic flow with fast diastolic drop in the common digital arteries. She had no other clinical manifestations of COVID-19. After receiving a therapeutic dose of low molecular weight heparin, her necrotic fingers were amputated without complications; histopathology revealed digital intravascular thrombosis." (Martino & Bitti 2020)

In San Francisco a patient who nearly died from Covid-19 had to have her fingers amputated as well. "Martha Macias, a cleaning professional and San Francisco Mission resident, describes the agonizing months of her fight against COVID-19. "Yo me enferme de Coronavirus y fue muy grave. Me cai en coma por un mes. Los doctors me decian que tenia el 10% de sobrevivir," said Macias. ("I had coronavirus, and it was terrible. It put me in coma for a month. Doctors told me I had a 10% chance of survival.") The coronavirus ravaged Macia's body for months. Doctors explained to her that the virus triggered a cluster of blood clots throughout her body. The lack of oxygen and blood flow threatened her legs and arms. She says doctors considered amputating both of her legs while she was in a coma."No es igual que me corten los dedos... que me corten la vida no," said Macias. ("It's not the same to get two fingers amputated to losing my life.")" (Pena 2020)

"The novel coronavirus SARS CoV-2 also appears to cause visible damage to the largest organ of the human body, the skin. There are reports from several countries that COVID-19 patients showed significant skin lesions. Small dermatological lesions on the feet have occurred particularly in children and young people. These purple patches resembled those caused by measles, chickenpox, or chilblains. On the toes, the lesions usually resembled frostbite or formed reticular patterns, normally caused by blood clots in small blood vessels. Sometimes, however, marks, redness, and hives-like rashes have also been observed on other parts of the body. It is possible that the bluish discoloration of the skin is due to pathological blood clotting, which could also be caused by the novel coronavirus." (Freund 2020)

Some patients have shown rare signs of Covid Tongue, and to be honest, we have seen some weird reactions to Covid-19 regarding this. A Houston mad was shown across all social media outlets with this overly swollen tongue, while others have shown discoloration and spots on the tongue that doctors have related to Covid-19. "When doctors studied 666 patients with Covid-19 in Spain, more than a tenth of them — 78 — exhibited "oral cavity findings," according to a study published in the British Journal of Dermatology. Of that group, 11 percent had inflammation of the small bumps on the tongue's surface; 6 percent had a swollen and inflamed tongue with indentations on the side; 6 percent had mouth ulcers; 4 percent had "patchy" areas on the tongue, and 4 percent had tissue swelling in the mouth. The oral cavity "deserves specific examination under appropriate circumstances to avoid contagion risk," the authors wrote." (Pawlowski 2021) Tongue features from a study done in china resemble elements of Covid tongue, which prompted researchers to look further into the matter. "Tongue features have certain relationship with the category of COVID-19. Tongue features can serve as potential indicators for the evaluation of patient's condition and prognosis. Further studies are needed to enhance the quantification of tongue features and develop standards." (Pang, Zhang, Lee 2020) As concluded in the study, more information is needed in regard to Covid Tongue. Is it a real thing? Yes, but the more we know, the better we can understand the impact Covid-19 is having on us all.

"Scientists around the world noticed an uptick in new diabetes cases last year and, in particular, saw that some COVID-19 patients with no history of diabetes were suddenly developing the condition, Scientific American reported. The trend prompted many research groups to launch studies of the phenomenon; for instance, researchers at King's College London in England and Monash University in Australia established the CoviDiab Registry, a resource where doctors can submit reports about patients with a confirmed history of COVID-19 and newly diagnosed diabetes. More than 350 clinicians have submitted reports to the registry, The Guardian reported. They've reported both type 1 diabetes, in which the body attacks the cells in the pancreas that produce insulin, and type 2 diabetes, in which the body still produces some insulin, though often not enough, and its cells don't respond properly to the hormone." (Lanese 2021)

We should be very alarmed about the uptick in Diabetes diagnosis, especially within our community. "The prevalence of COVID-19-associated diabetes is not the result of a single event but of a combination of disease susceptibility associated with chronic illness and COVID-19-specific mechanisms affecting metabolism. Whether a separate entity of post-COVID-19 diabetes, possibly associated with lasting β-cell damage, also exists is not yet clear." (Accili 2021)

"For some people, surviving COVID-19 may lead to lasting medical concerns, including newly diagnosed diabetes. This chronic disease happens when your blood glucose, also called blood sugar, is too high. The pancreas makes insulin, which is a hormone that helps glucose get into your cells to be used for energy. But when your body doesn't make enough—or any—insulin (Type 1 diabetes) or doesn't use insulin well (Type 2 diabetes), glucose then stays in your blood and can cause health problems. In November 2020, a global analysis published in the journal Diabetes, Obesity and Metabolism found that up to 14.4% of people who were hospitalized with severe COVID-19 also developed diabetes." (Wyne M.D. Ph.D. 2021)

Until earlier this year, I had only read a headline associated with a claim that Covid-19 potentially cause Type I Diabetes; however, further studies are currently on the way to help explain the relationship between Covid-19 and Diabetes.

What Sars-Cov-2 Taught Me was more than just a virus that seeks out a host. Its novel presence came with a disease (Covid-19) that preyed on people in ways other respiratory illnesses never had. As stated above, from head to toe, we see evidence of the damage left behind after infection and the lingering effects that exist post-Covid.

"No part of our body is safe from this disease; even if we recover from a natural infection, we still may have lingering health effects that will exist. Post-COVID conditions are a wide range of new, returning, or ongoing health problems people can experience more than four weeks after first being infected with the virus that causes COVID-19. Even people who did not have symptoms when they were infected can have post-COVID conditions. These conditions can have different types and combinations of health problems for different lengths of time." (CDC 2021) As we approach year 3, information will come forth especially in the field of long covid.

"Much is still unknown about how COVID-19 will affect people over time, but research is ongoing. Researchers recommend that doctors closely monitor people who have had COVID-19 to see how their organs are functioning after recovery. Many large medical centers are opening specialized clinics to provide care for people with persistent symptoms or related illnesses after recovering from COVID-19. Support groups are available as well. It's important to remember that most people who have COVID-19 recover quickly. But the potentially long-lasting problems from COVID-19 make it even more important to reduce the spread of COVID-19 by following precautions. Precautions include wearing masks, social distancing, avoiding crowds, getting a vaccine when available, and keeping hands clean." (Mayo Clinic 2021)

Even with taking these precautionary measures, we have to be aware of the many different variants that exist, which could lead to breakthrough infection and may increase risk of potential threat of disease. This is 'What Sars-Cov2 Taught Me' and to add, "Novel coronavirus (COVID-19) symptoms can last weeks or months for some people. These patients, given the name "long haulers", have in theory recovered from the worst impacts of COVID-19 and have tested negative. However, they still have symptoms. There seems to be no consistent reason for this to happen. Researchers estimate about 10% of COVID-19 patients become long haulers, according to a recent article from The Journal of the American Medical Association and a study done by British scientists. That's in line with what UC Davis Health is seeing. This condition can affect anyone –old and young, otherwise healthy people and those battling other conditions. It has been seen in those who were hospitalized with COVID-19 and patients with very mild symptoms." (UC Davis Health 2021) Be safe!!!!!!!

Long COVID Symptoms (additional information)

Long COVID is a range of symptoms that can last weeks or months after first being infected with the virus that causes COVID-19 or can appear weeks after infection. Long COVID can happen to anyone who has had COVID-19, even if the illness was mild, or they had no symptoms. People with long COVID report experiencing different combinations of the following symptoms:

- Tiredness or fatigue
- Difficulty thinking or concentrating (sometimes referred to as "brain fog")
- Headache
- Loss of smell or taste
- Dizziness on standing
- Fast-beating or pounding heart (also known as heart palpitations)
- Chest pain
- Difficulty breathing or shortness of breath
- Cough
- Joint or muscle pain
- Depression or anxiety
- Fever
- Symptoms that get worse after physical or mental activities

(National Center for Immunization and Respiratory Diseases (NCIRD), Division of Viral Diseases 2021)

Chapter IV
Review

1). What name has been given to patients that have recovered from Covid-19?

2). Can you name some of the symptoms Long Covid patients experience after being infected with the virus?

3). What is Covid Toes?

4). _____ can lead to rashes in weird places and in some instances bumps along the armpits hands and leg area.

5). True or False: Covid fingers are similar to skin rashes.

6). Is Covid-19 a contributing factor to new cases of Diabetes?

7). What organs are impacted by Covid-19?

8). True or False: Two recent studies in the journal Nature provide some of the most-detailed analyses yet about the effects on the human body of SARS-CoV-2, the coronavirus that causes COVID-19.

9). What did researchers grow in order to learn more about how Covid impacts the body?

10). SARS-CoV-2 destroy cells within air sacs called?

11). True or False: Covid-19 can cause brain fog.

Chapter IV Sources

- Newman, Tim. "Links between COVID-19 and Skin Rashes." Translated by Alexandra Saffins, Medical News Today, MediLexicon International, 18 Mar. 2021, www.medicalnewstoday.com/articles/links-between-covid-19-and-skin-rashes.

- Wyne, Kathleen. "Why Are People Developing Diabetes after Having COVID-19?" Ohio State Medical Center, 19 Feb. 2021, wexnermedical.osu.edu/blog/why-are-people-developing-diabetes-after-having-covid19.

- Clinic, Mayo. "COVID-19 (Coronavirus): Long-Term Effects." Mayo Clinic, Mayo Foundation for Medical Education and Research, 6 May 2021, www.mayoclinic.org/diseases-conditions/coronavirus/in-depth/coronavirus-long-term-effects/art-20490351#:~:text=Even%20in%20young%20people%2C%20COVID,Parkinson's%20disease%20and%20Alzheimer's%20disease.

- Division of Viral Diseases, National Center for Immunization and Respiratory Diseases (NCIRD). "Post-COVID Conditions." Centers for Disease Control and Prevention, Centers for Disease Control and Prevention, 8 Apr. 2021, www.cdc.gov/coronavirus/2019-ncov/long-term-effects.html.

- Health, UC Davis. "Long Haulers: Why Some People Experience Long-Term Coronavirus Symptoms." Long Haulers Suffer Long-Term Coronavirus Symptoms | UC Davis Health, 8 Feb. 2021, health.ucdavis.edu/coronavirus/covid-19-information/covid-19-long-haulers.html.

- Assoc, American Lung. "Learn about COVID-19." Learn about COVID-19 | American Lung Association, American Lung Association Scientific and Medical Editorial Review Panel., 14 Apr. 2021, www.lung.org/lung-health-diseases/lung-disease-lookup/covid-19/about-covid-19.

- Collins, Francis. "How Severe COVID-19 Can Tragically Lead to Lung Failure and Death." National Institutes of Health, U.S. Department of Health and Human Services, 10 May 2021, directorsblog.nih.gov/2021/05/11/how-severe-covid-19-can-tragically-lead-to-lung-failure-and-death/.

- Post, Wendy S. "Heart Problems after COVID-19." Johns Hopkins Medicine, 2021, www.hopkinsmedicine.org/health/conditions-and-diseases/coronavirus/heart-problems-after-covid19.

- Assoc, AAD. "COVID Toes, Rashes: How the Coronavirus Can Affect Your Skin." American Academy of Dermatology, 2021, www.aad.org/public/diseases/coronavirus/covid-toes.

- *Bharat, MBBS , Prof Ankit, et al. Early Outcomes after Lung Transplantation for Severe COVID-19: a Series of the First Consecutive Cases from Four Countries, The Lancent , 31 Mar. 2021, www.thelancet.com/journals/lanres/article/PIIS2213-2600(21)00077-1/fulltext.*

- Bharat, Ankit, et al. "Lung Transplantation for Patients with Severe COVID-19." Science Translational Medicine, American Association for the Advancement of Science, 16 Dec. 2020, stm.sciencemag.org/content/12/574/eabe4282.

- Weir, Kristen. "How COVID-19 Attacks the Brain." Monitor on Psychology, American Psychological Association, 11 Nov. 2020, www.apa.org/monitor/2020/11/attacks-brain#:~:text=In%20a%20review%20of%20case,in%20which%20the%20immune%20system.

- Hamilton, Jon. "How COVID-19 Attacks The Brain And May Cause Lasting Damage." NPR, NPR, 5 Jan. 2021, www.npr.org/sections/health-shots/2021/01/05/953705563/how-covid-19-attacks-the-brain-and-may-cause-lasting-damage.

- Shimabukuro and Others, T.T., et al. "Microvascular Injury in the Brains of Patients with Covid-19: NEJM." New England Journal of Medicine, 21 Apr. 2021, www.nejm.org/doi/full/10.1056/NEJMc2033369.

- University, Georgia State. "Study Finds COVID-19 Attack on Brain, Not Lungs, Triggers Severe Disease in Mice." ScienceDaily, ScienceDaily, 19 Jan. 2021, www.sciencedaily.com/releases/2021/01/210119114456.htm.

- Caffrey, Mary. "Viral RNA Found in Kidneys of COVID-19 Patients." AJMC, AJMC, 19 Aug. 2020, www.ajmc.com/view/lancet-covid-19-virus-can-attack-kidneys-speeding-death.

- Budson, Andrew. "What Is COVID-19 Brain Fog - and How Can You Clear It?" Harvard Health, 8 Mar. 2021, www.health.harvard.edu/blog/what-is-covid-19-brain-fog-and-how-can-you-clear-it-2021030822076.

- Fruend, Alexander. "How the Novel Coronavirus Attacks Our Entire Body: DW: 11.05.2020." DW.COM, Deutsche Welle, 5 Nov. 2020, www.dw.com/en/how-the-novel-coronavirus-attacks-our-entire-body/a-53389908.

- Madjid, Mohammad. "Potential Effects of Coronaviruses on the Cardiovascular System." JAMA Cardiology, JAMA Network, 1 July 2020, jamanetwork.com/journals/jamacardiology/fullarticle/2763846.

- Bleicher, Ariel, and Katherine Conrad. "We Thought It Was Just a Respiratory Virus." We Thought It Was Just a Respiratory Virus | UC San Francisco, University of California San Francisco, 10 June 2021, www.ucsf.edu/magazine/covid-body.

- Mallapaty, Smriti. "Mini Organs Reveal How the Coronavirus Ravages the Body." Nature News, Nature Publishing Group, 22 June 2020, www.nature.com/articles/d41586-020-01864-x.

- Marjot, Thomas, et al. "COVID-19 and Liver Disease: Mechanistic and Clinical Perspectives." Nature News, Nature Publishing Group, 10 Mar. 2021, www.nature.com/articles/s41575-021-00426-4.

- Pena, Luz. "SF Woman Who Will Have Fingers Amputated after Nearly Dying from COVID-19, Still Hesitant about Vaccine." ABC7 San Francisco, KGO-TV, 18 Dec. 2020, abc7news.com/coronavirus-latino-covid-rate-fingers-amputated-vaccine-news/8784805/.

- Martino, Giuseppe P, and Giuseppina Bitti. "DEFINE_ME." COVID Fingers: Another Severe Vascular Manifestation, European Society for Vascular Surgery. Published by Elsevier B.V. , 14 Aug. 2020, www.ejves.com/article/S1078-5884(20)30676-6/fulltext.

- Shimabukuro and Others, T.T., et al. "New-Onset Diabetes in Covid-19: NEJM." New England Journal of Medicine, 21 Apr. 2021, www.nejm.org/doi/full/10.1056/nejmc2018688.

- Lanese, Nicoletta. "COVID-19 May Trigger Diabetes in Some People." LiveScience, Purch, 22 Mar. 2021, www.livescience.com/covid19-may-trigger-diabetes.html.

- Accili, Domenico. "Can COVID-19 Cause Diabetes?" Nature News, Nature Publishing Group, 11 Jan. 2021, www.nature.com/articles/s42255-020-00339-7.

- Pawlowski, A. "'Covid Tongue' May Be Another Coronavirus Symptom, British Researcher Suggests." NBCNews.com, NBCUniversal News Group, 29 Jan. 2021, www.nbcnews.com/health/health-news/covid-tongue-may-be-another-coronavirus-symptom-british-researcher-suggests-n1256078.

- Pang, Wentai, et al. "Tongue Features of Patients with Coronavirus Disease 2019: a Retrospective Cross-Sectional Study." Integrative Medicine Research, Elsevier, 25 July 2020, www.sciencedirect.com/science/article/pii/S2213422020301256.

The Evolution of Vaccine Science & The Creation of Covid-19 mRNA Vaccines
Chapter V

The history of vaccines is well documented, and since the early instances of inoculations, people have had issues with this medical solution to a widespread problem. The religious community deemed it witchcraft evil and against the will of God. God is cultural everyone has some form of a god deity that is totally different from the monotheistic gods, and for one to deem a practice evil and witchcraft was in a sense disrespectful to other cultural ideologies. However, inoculations evolved in many instances independent of one's cultural worldview; it is not exclusive to Africa, China, or India. This shows the genius of homo-sapiens sapiens looking to resolve things that affected our existence.

Viruses have plagued mankind since history has recorded such instances. "Evidence for knowledge of several diseases that we know now to be caused by viruses can be found in ancient records. The Greek poet Homer characterizes Hector as 'rabid' in The Iliad, and Mesopotamian laws that outline the responsibilities of the owners of rabid dogs' date from 1000 BCE. Their existence indicates that the communicable nature of this viral disease was already well-known by that time. Egyptian hieroglyphs illustrate what appear to be consequences of poliovirus infection (a withered leg typical of poliomyelitis). Pustular lesions characteristic of smallpox has also been found in Egyptian mummies. The smallpox virus was probably endemic in the Ganges River basin by the fifth century BCE and subsequently spread to other parts of Asia and Europe. This viral pathogen has played an important part in human history. Its introduction into the previously unexposed native populations of Central and South America by colonists in the 16th century led to lethal epidemics, which are considered an important factor in the conquests achieved by a small number of European soldiers. Other viral diseases known in ancient times include mumps and, perhaps, influenza. Europeans have described yellow fever since they discovered Africa, and it has been suggested that this scourge of the tropical trade was the basis for legends about ghost ships' such as the Flying Dutchman, in which an entire ship's crew perished mysteriously." (Flint, Racaniello, Rall, Hatziioannou, andSkalka 2020:7)

In short certain viruses and human interactions are very familiar with one another however, prevention to specific pathogens across the globe indicates that some viruses won, and humans lost. But not in all instances because human manipulation of viruses became known among certain cultural practices that lead to solutions to the problem. "Inoculation for smallpox, the forerunner of vaccination, has had a long and variegated history. The earliest known descriptions of the practice in sub-Saharan Africa were given by African slaves in colonial America in the early and mid-eighteenth century. Subsequently, it is mentioned in accounts from widely scattered parts of the continent. It seems to have been most extensively used in the Western and Central Sudan, Ethiopia, and Southern Africa. Local diffusion patterns emerge from the available evidence, but broader questions of origin must await further investigation. Similarly, it is yet impossible to assess its demographic impact in Africa although it clearly provided some defense against smallpox despite the risks involved." (Herbert 2009)

What Africans did to resolve their smallpox issues at that time was genius but dangerous. By making an incision in the skin and introducing the body to a foreign pathogen allowed the virus to be naturally introduced to the body. After several days immunity would be gained; this practice would travel as stated above to the new world during the middle of a transcontinental holocaust, otherwise known as the slave trade. One familiar story about the introduction to inoculation in the Western world Onesimus (an intelligent fellow) who was purchased by Cotton Mather in 1706 a slave owner who had asked Onesimus if he knew anything about smallpox. His answer was yes and no, and he explained he had undergone a procedure that provided him protection against such a virus. Mather would continue to ask other Africans if this was true for them, and they confirmed such treatment. This was breaking news at this time because Mather would take this information to the Royal Society a few years later with hopes of increasing the value of slaves who have been inoculated against smallpox.

This was on the cusp of 1721's Boston outbreak of smallpox. From this moment forth, the evolution of inoculation continued by using lesions on the skin and scratching healthy people with a lancet that contained a portion of the virus. "In the 1790s, Edward Jenner, an English country physician, established the principle on which modern methods of viral immunization are based, even though viruses themselves were not to be identified for another 100 years. As a young boy, Jenner himself was variegated with smallpox and was undoubtedly familiar with its effects and risks. Perhaps this experience spurred his abiding interest in this method. Although it is commonly asserted that Jenner's development of the smallpox vaccine was inspired by his observation of milkmaids, the reality is more prosaic. As a physician's apprentice at age 13, Jenner learned about a curious observation of local practitioners who had been variegating farmers with smallpox. No expected skin rash or disease appeared in farmers who had previously suffered a bout with cowpox. This lack of response was typical of individuals who had survived earlier injections with smallpox and were known to be immune to the disease. It was supposed, therefore, that, like smallpox survivors, these non-responding farmers must the first observed and later reported by others; Jenner was the first to appreciate its significance fully and to follow up with direct experiments. From 1794 to 1796, he demonstrated that inoculation with material from cowpox lessons induced only mild symptoms in the recipient but protected against the far more dangerous disease. It is from these experiments that we derive the term vaccination (vacca= 'cow' in Latin); Louis Pasteur coined this term in 1881 to honor Jenner's accomplishments." (Flint, Racaniello, Rall, Hatziioannou, and Skalka 2020:9)

Here we have two examples of the early history of inoculations/vaccinations in human history, which set the stages for 1st generation vaccines that held a risk-reward in its early history. "Initially, the only way to propagate and maintain the cowpox-derived vaccine was by serial infection of human subjects. This method was eventually banned, as it was often associated with transmission of other diseases such as syphilis and hepatitis. By 1860, the vaccine had been passaged in cows; later, horses, sheep, and water buffaloes were also used. The origin of the current vaccine virus, vaccinia virus, is now thought to be horse-pox virus." (Flint, Racaniello, Rall, Hatziioannou, and Skalka 2020:9)

In the earlier onset of 1st generation vaccines and resolving different viruses that were known to man, the vaccine-making process could take numerous years to develop. "The main goal of a vaccine for a particular infectious agent, such as the virus that causes COVID-19, is to teach the immune system what that virus looks like. Once educated, the immune system will vigorously attack the actual virus if it ever enters the body. Viruses contain a core of genes made of DNA or RNA wrapped in a coat of proteins. To make the coat of protein, the DNA or RNA genes of the virus make messenger RNA (mRNA); the mRNA then makes the proteins. An mRNA of a specific structure makes a protein of a specific structure." (Komaroff MD 2020)

1st generation vaccines are: "Attenuated and inactivated vaccines are identified in the first generation, which uses a primary method in their production. Attenuated pathogens, full organisms, or inactivated bacterial toxins, which are effectively immunogenic, are used in making these vaccines. There are some advantages in these kinds of vaccines due to their high ability to stimulate innate immunity, induction of long-term protection, easy production, and low production costs. However, this generation has some disadvantages, such as inducing disease due to the use of complete pathogen (live or inactivated) and virulence recursively of the pathogen in the host body. This type of vaccine is known as a traditional vaccine." (Tahamtan, Charostad, Barati, Shokouh 2017)

For instance, the next antiviral vaccine known to us as the yellow fever took nearly 50 years to create. Partly because they wanted to make vaccines safer and more efficient but mostly be able to produce vaccines in large quantities. The late mid to late 1800s and early 1900s was a period of scientific gain. The first virus had been seen under a microscope, Darwin's success with Natural Selection and Leeuwenhoek's understanding of microorganisms helps drive fierce debates, but the instrument that revolutionized virology, "The Electron Microscope," provided answers to questions virologists had.

2nd generation vaccines wouldn't be discovered until the 1960s that were called adenovirus vaccines. Adenovirus vaccines are: "nonenveloped, double-stranded DNA viruses (genome size: 34-43 kb; virion size: 70-90 nm), first discovered in the human adenoid tissue in 1953 by Rowe and his colleagues. In humans, adenoviruses generally cause mild respiratory and gastrointestinal tract infections; however, adenovirus-induced infections can be life-threatening in immunocompromised people or people with pre-existing respiratory or cardiac disorders. These viruses are isolated from various mammalian species, ranging from simians to chimpanzees to human beings. In humans, more than 50 adenovirus serotypes (no cross-neutralization by antibodies) have been identified, which are divided into 7 subgroups (A – G) based on red blood cell agglutination properties and sequence homology. Adenoviruses express two types of genes: early genes and late genes. Early genes (E1A, E1B, E2, E3, and E4) are necessary for supporting viral replication inside host cells, whereas late genes are required for host cell lysis, viral assembly, and virion release. Recombinant adenoviruses that are generated in the laboratory as vectors can be either replication-deficient or replication-competent. Because the E1 gene is essential for viral replication, experimental depletion of the E1 gene generates adenoviruses that are capable of infecting the host cells but cannot grow in numbers because of defective replication. However, some specialized cells, such as HEK 293, can facilitate the replication of E1-deficient adenoviruses by providing E1 functions in trans." (Dutta, Ph.D. 2021)

Manufacturing 2nd generation vaccines reduced the amount of time it took to make first-generation vaccines. This feat was evolutionary and science-supported this feat as it helped resolve aged old viruses that once plagued humanity. "Today, the rise of genetic engineering and molecular biology has greatly impacted the development and manufacturing process of vaccines. Specific antigenic microbes have high power to arouse the immune response against pathogens. Currently, the sequence of the pathogenic protein antigens could be obtainable by sequencing genes of the main antigen and producing them synthetically via recombinant DNA technology. Hepatitis B is the first and one of the most successful examples of synthetic vaccines. The surface antigen of this virus (HBsAg) is very immunogenic and effective and can produce high antibody levels in the body. In the past, for providing hepatitis B vaccine, HBsAg was purified from the plasma of infection carriers and used for vaccination; of course, there were some extensive restrictions in purification, such as difficult conditions and contaminated plasma. In order to make recombinant hepatitis B vaccine, recombinant HBsAg is expressed in cells that have a powerful expression system (such as yeast), leading to the production of virus-like particles by HBsAgs, which are highly immunogenic. Since these particles have no genome, they do not create disease and lead to effective and powerful responses against the main pathogen. Other kinds of common vaccines are anti-herpes simplex virus, anti-rotavirus, and anti-HPV vaccines." (Tahamtan, Charostad, Barati, Shokouh 2017)

2nd Generation Vaccines orchestrated change that people appreciated because it improved its safety and efficacy over time and provided a pathway to allow technology to help it improve in areas that would help resolve issues faster. Anti-vaxxers complained about vaccine ingredients related to certain vaccines. In some instances, anti-vaxxers would lump vaccine ingredients together and argue that they all contained eggs, fetuses, and mercury. Also, these vaccines were known to have efficacious rates above 50%, which is the threshold allowed. "Traditional vaccines work: polio and measles are just two examples of serious illnesses brought under control by vaccines. Collectively, vaccines may have done better for humanity than any other medical advance in history. But growing large amounts of a virus, and then weakening the virus or extracting the critical piece, takes a lot of time." (Komaroff MD 2020) However, adenoviruses are limited in their ability to be readily available in a timely manner in the case of the current pandemic.

56 years later, the emergence of 3rd Generation Vaccines would evolve by chance due to a novel virus that spread across the globe faster than Chris Johnson of the Tennessee Titans did on his way to the end zone from 90 plus yards out. "When Katalin Karikó, Ph.D., came to the United States from Hungary in 1985, she brought with her a passionate determination to work on mRNA. Messenger RNA is fundamental to life sets of blueprints, spelled out using four nucleotides "letters," for building every protein in every life form on Earth. Karik'ós big idea was to produce proteins at will by injecting mRNA into cells, but her experiments did not work for a long time. Lack of success forced her to rely on one senior scientist after another to support her work, while she made only meager wages. In 1998, Karikó partnered with Drew Weissman, M.D., Ph.D., at the University of Pennsylvania. Weissman was interested in developing an HIV vaccine based on mRNA. After many failures, Karikó and Weissman learned that natural mRNAs use small amounts of slightly modified nucleotides, in addition to the four standard nucleotides. When the scientists inserted the modified nucleotides into the mRNAs they were using in their research; they began to find that these modified mRNAs produced proteins efficiently without causing undesirable side effects. They began to publish their findings, starting in 2005. By the time the coronavirus that causes COVID-19 showed up in 2020, Karikó and Weissman were already working on an influenza vaccine based on their mRNA technology." (Nahm, M.D. 2021)

3rd Generation Vaccines are: "Immunogenic potential administration of a plasmid containing a gene encoding the antigen, known as genetic vaccines, is categorized as third generation vaccines, and is a valuable method, which researchers have considered since the beginning of 1990s. Different names have been given for this kind of vaccine, such as DNA vaccines, RNA vaccines, and plasmid vaccines. The expertise committee of WHO vaccination in 1996 chose nucleotide acid that includes both DNA and RNA vaccines. Furthermore, genetic i/mmunization and DNA immunization terms were used for this type of immunization." (Tahamtan, Charostad, Barati, Shokouh 2017)

The first of 3rd generation vaccines are mRNA vaccines. Research surrounding 3rd generation vaccines began 30 years ago and what scientist have learned recently is that it's easy to make mRNA in a laboratory setting, but mRNA vaccines could not get over its hump during critical testing that slowed its emergence from being available much sooner. With the emergence of a novel virus, scientist had another opportunity to utilize their research and tackle a global problem. "Then along came COVID-19. Within weeks of identifying the responsible virus, scientists in China had determined the structure of all of its genes, including the genes that make the spike protein, and published this information on the Internet. Within minutes, scientists 10,000 miles away began working on the design of an mRNA vaccine. Within weeks, they had made enough vaccine to test it in animals and then in people. Just 11 months after the discovery of the SARS-CoV-2 virus, regulators in the United Kingdom and the US confirmed that an mRNA vaccine for COVID-19 is effective and safely tolerated, paving the path to widespread immunization. Previously, no new vaccine had been developed in less than four years." (Komaroff MD 2020) mRNA vaccines are currently tackling issues such as Ebola, certain Cancers, and HIV, just to name a few. "mRNA vaccines have been studied before for flu, Zika, rabies, and cytomegalovirus (CMV). As soon as the necessary information about the virus that causes COVID-19 was available, scientists began designing the mRNA instructions for cells to build the unique spike protein into an mRNA vaccine. Future mRNA vaccine technology may allow for one vaccine to provide protection for multiple diseases, thus decreasing the number of shots needed for protection against common vaccine-preventable diseases." (National Center for Immunization and Respiratory Diseases (NCIRD), Division of Viral Diseases 2021)

"In order to produce vaccines against COVID-19, Moderna and Pfizer/BioNTech first needed to identify the right mRNA sequences for proteins that would make suitable vaccine components. COVID-19 virus is similar to the SARS virus of 2003, and previous research of the SARS outbreak suggested potential vaccine components. Nevertheless, exact design of the actual vaccine structure was not a simple task for a small company like Moderna, or even a larger one such as Pfizer, even though the entire COVID-19 virus genome had been published by Chinese researchers in early 2020. But here is where another decades-long scientific effort paid off: Concern about another influenza pandemic on the scale of the 1918 flu pandemic has long motivated U.S. investment in flu studies, particularly by the NIH. The NIH has also spent more than 40 years funding scientists working toward a vaccine for HIV. The crucial sequence information needed to target COVID-19 came from a vaccine research laboratory at the NIH. The lab had been working on HIV vaccine designs for decades and had developed algorithms to identify molecular structures optimized for vaccines. This approach also predicted a successful vaccine structure against the respiratory syncytial virus. Harnessing that experience, the NIH group rapidly identified stable molecular structures optimized for the mRNA vaccines against COVID-19." (Nahm, M.D. 2021)

"Some vaccines use a whole virus or bacterium to teach our bodies how to build up immunity to the pathogen. These pathogens are inactivated or attenuated, which means weakened. Other vaccines use parts of viruses or bacteria. Recombinant vaccine technology employs yeast or bacterial cells to make many copies of a particular viral or bacterial protein or sometimes a small part of the protein. mRNA vaccines bypass this step. They are chemically synthesized without the need for cells or pathogens, making the production process simpler. mRNA vaccines carry the information that allows our own cells to make the pathogen's proteins or protein fragments themselves. Importantly, mRNA vaccines only carry the information to make a small part of a pathogen. From this information, it is not possible for our cells to make the whole pathogen." (Martin, Ph.D. 2020) This science is critical in helping the body build its own spike protein to fight against the virus. Every part of the cell plays an intricate role, and this technology has saved millions of lives as of today. "Vaccines work because they educate the host's immune system to recall the identity of a specific virus years after the initial encounter, a phenomenon called immune memory. The resounding practical success of immunization in stimulating long-lived immune memory is among humanity's greatest scientific and medical advancements." (Flint, Racaniello, Rall, Hatziioannou, and Skalka 2020:239)

"RNA is a notoriously fragile molecule. Delivering mRNA successfully to cells inside our bodies and ensuring that enzymes within our cells do not degrade it are key challenges in vaccine development. Chemical modifications during the manufacturing process can significantly improve the stability of mRNA vaccines. Encapsulating mRNA in lipid nanoparticles is one way to ensure that a vaccine can successfully enter cells and deliver the mRNA into the cytoplasm. mRNA does not linger in our cells for long. Once it has passed its instructions to the protein-making machinery in our cells, enzymes called ribonucleases (RNases) degrade the mRNA. It is not possible for mRNA to move into the nucleus of a cell as it lacks the signals that would allow it to enter this compartment. This means that RNA cannot integrate into the DNA of the vaccinated cell. There is no risk of long-term genetic changes with mRNA vaccines. The mRNA COVID-19 vaccines by Pfizer and Moderna have undergone safety testing in human clinical trials." (Martin, Ph.D. 2020)

These are some of the things that Sars-CoV2 taught me over the past year. It allowed me to understand how the mRNA vaccine works, and once I understood that I focused on the scientific data, primarily the Randomized Controlled Trials that Moderna, Pfizer, and Johnson & Johnson have all been approved (Emergency Use Authorization) for. As of lately, both Moderna and Pfizer have both applied for BLA or full authorization. One of the major questions asked in the community is are the vaccines safe. Saying it's safe without evidence holds absolutely no weight among people, but I can say it is not safe without evidence, and people will be very receptive of that and did people die from taken the vaccine in the next chapter, we will review some of those underlining questions and claims made by vaccine-hesitant people will be address at that time. However, we will review the safety data as it was submitted to the FDA.

Moderna's Safety Data

The information provided by the Sponsor was adequate for review and to make conclusions about the safety of the mRNA-1273 vaccine in the context of the proposed indication and population for intended use under EUA. The number of participants in the Phase 3 safety population (N=30,350; 15,184 vaccine,15,165 placebo) meets the expectations described in FDA's Guidance on Development and Licensure of Vaccines to Prevent COVID-19 for efficacy. The initial EUA request was based on data from the pre-specified interim analysis (November 11, 2020, data cutoff) with a median follow-up duration of 7 weeks after dose 2; this interim analysis data is the primary basis of this EUA review and conclusions. Data and analyses from a November 25, 2020, data cut with a median duration of at least 2 months follow-up after completion of the 2-dose primary vaccination series was submitted as an amendment to the EUA request on December 7, 2020. The FDA has not independently verified the complete safety data from the primary analysis, aside from all new deaths (including those reported through December 3, 2020) and SAEs. No new safety concerns have been identified. The rates and types of solicited adverse reactions and unsolicited adverse events are unlikely to change significantly with an additional 2 weeks of follow-up. The totality of the data package submitted in the EUA request meets the Agency's expectations on the minimum duration of follow-up.

Local site reactions and systemic solicited events after vaccination were frequent and mostly mild to moderate. The most common solicited adverse reactions were injection site pain (91.6%), fatigue (68.5%), headache (63.0%), muscle pain (59.6%), joint pain (44.8%), and chills (43.4%); 0.2% to 9.7% were reported as severe, with severe solicited adverse reactions being more frequent after dose 2 than after dose 1 and generally less frequent in adults ≥65 years of age as compared to younger participants. Among adverse events of clinical interest, lymphadenopathy was reported in 173 participants (1.14%) in the vaccine group and 95 participants (0.63%) in the placebo group. There was a numerical imbalance in hypersensitivity adverse events across study groups, with 1.5% of vaccine recipients and 1.1% of placebo recipients reporting such events in the Safety Set. There were no anaphylactic or severe hypersensitivity reactions with close temporal relation to the vaccine.

Throughout the safety follow-up period to date, there have been three reports of Bell's palsy in the vaccine group and one in the placebo group. Currently, available information is insufficient to determine a causal relationship with the vaccine. There were no other notable patterns or numerical imbalances between treatment groups for specific categories of adverse events (including other neurologic, neuro-inflammatory, and thrombotic events) that would suggest a causal relationship to mRNA- As of December 3, 2020, there were a total of 13 deaths reported in the study (6 vaccine, 7 placebo). These deaths represent events and rates that occur in the general population of individuals in these age groups. The frequency of non-fatal serious adverse events was low and without meaningful imbalances between study arms (1% in the mRNA-1273 group and 1% in the placebo group). The most common SAEs in the vaccine group, which were numerically higher than the placebo group, were myocardial infarction (0.03%), cholecystitis (0.02%), and nephrolithiasis (0.02%), although the small numbers of cases of these events do not suggest a causal relationship. The most common SAEs in the placebo arm, which were numerically higher than the vaccine arm, aside from COVID-19 (0.1%), were pneumonia (0.05%) and pulmonary embolism (0.03%).

(FDA Briefing Document Moderna COVID-19 Vaccine 2020)

Pfizer Safety Data

The information provided by the Sponsor was adequate for review and to make conclusions about the safety of BNT162b2 in the context of the proposed indication and population for intended use under EUA. The number of participants in the phase 2/3 safety population (N=37586; 18801 vaccine,18785 placebo) meets the expectations in FDA's Guidance on Development and Licensure of Vaccines to Prevent COVID-19 for efficacy, and the median duration of at least 2 months follow-up after completion of the 2-dose primary vaccination series meets the agency's expectations in FDA's Guidance on its Emergency Use Authorization for Vaccines to Prevent COVID-19. The all-enrolled population contained more participants >16 years of age, regardless of the duration of follow-up (43448; 21720 vaccines, 21728 placeboes). The demographic and baseline characteristics of the all-enrolled population and the safety population were similar. Although the overall median duration of follow-up in the all-enrolled population was less than 2 months, because the protocol was amended to include subpopulations such as individuals with HIV and adolescents, the data from both populations altogether provide a comprehensive summary of safety. Local site reactions and systemic solicited events after vaccination were frequent and mostly mild to moderate. The most common solicited adverse reactions were injection site reactions (84.1%), fatigue (62.9%), headache (55.1%), muscle pain (38.3%), chills (31.9%), joint pain (23.6%), fever (14.2%); severe adverse reactions occurred in 0.0% to 4.6% of participants, were more frequent after Dose 2 than after Dose 1 and were generally less frequent in adults ≥55 years of age (≤2.8%) as compared to younger participants (≤4.6%). Among adverse events of special interest, which could be possibly related to vaccine, lymphadenopathy was reported in 64 participants (0.3%): 54 (0.5%) in the younger (16 to 55 years) age group; 10 (0.1%) in the older (>55 years) age group; and 6 in the placebo group. The average duration of these events was approximately 10 days, with 11 events ongoing at the time of the data cutoff. Bell's palsy was reported by four vaccine participants. From Dose 1 through 1 month after Dose 2, there were three reports of Bell's palsy in the vaccine group and none in the placebo group. This observed frequency of reported Bell's palsy is consistent with the expected background rate in the general population. There were no other notable patterns or numerical imbalances between treatment groups for specific categories of non-serious adverse events (including other neurologic, neuro-inflammatory, and thrombotic events) that would suggest a causal relationship to BNT162b2 vaccine. A total of six deaths occurred in the reporting period (2 deaths in the vaccine group, 4 in placebo). In the vaccine group, one participant with baseline obesity and pre-existing atherosclerosis died 3 days after Dose 1, and the other participant experienced cardiac arrest 60 days after Dose 2 and died 3 days later. Of the four deaths in the placebo arm, the cause was unknown for two of them, and the other two participants died from hemorrhagic stroke (n=1) and myocardial infarction (n=1), respectively; three deaths occurred in the older group (>55 years of age). All deaths represent events that occur in the general population of the age groups where they occurred at a similar rate. The frequency of non-fatal serious adverse events was low (<0.5%), without meaningful imbalances between study arms. The most common SAEs in the vaccine arm which were numerically higher than in the placebo arm were appendicitis (0.04%), acute myocardial infarction (0.02%), and cerebrovascular accident (0.02%), and in the placebo arm numerically higher than in the vaccine arm were pneumonia (0.03%), atrial fibrillation (0.02%), atrial fibrillation (0.02%) and syncope (0.02%). Appendicitis was the most common SAE in the vaccine arm. There were 12 participants with SAEs of appendicitis: 8 in the BNT162b2 group. Of the 8 total appendicitis cases in the BNT162b2 group, 6 occurred in the younger (16 to 55 years) age group and 2 occurred in the older (>55 years) age group (one of the cases in the older age group was perforated). One of the 6 participants with appendicitis in the younger age group also had a peritoneal abscess. Cases of appendicitis in the vaccine group were not more frequent than expected in the general population.

(FDA Briefing Document Pfizer-BioNTech COVID-19 Vaccine 2020)

Johnson & Johnson's Janseen Safety Data

The information provided by the Sponsor was adequate for review and to make conclusions about the safety of the Ad26.COV2.S vaccine in the context of the proposed indication and population for the intended use under EUA. The number of participants in the Phase 3 safety population (N=43,783; 21,895 vaccine, 21,888 placeboes) meets the expectations for efficacy in FDA's guidance for industry Development and Licensure of Vaccines to Prevent COVID-19 (June 2020). A subset of participants (N=6,736) was followed for solicited reactions within 7 days following vaccination and unsolicited reactions within 28 days following vaccination. The demographic and baseline characteristics of the all-enrolled population and the safety subset were similar with respect to age and sex but had imbalances with respect to race, baseline comorbidities, SARS-CoV-2 serostatus and geographic distribution. Local site reactions and systemic solicited events among vaccine recipients were frequent and mostly mild to moderate. The most common solicited adverse reactions were injection site pain (48.6%), headache (38.9%), fatigue (38.2%) and myalgia (33.2%); 0.7% and 1.8% of local and systemic solicited adverse reactions, respectively, were reported as grade 3. Overall, solicited reactions were reported more commonly in younger participants. There were no meaningful imbalances in unsolicited adverse events in 28 days following vaccination between vaccine and placebo recipients in the safety subset. Among all adverse events collected through the data cutoff of January 22, 2021, a numerical imbalance was seen in urticaria events reported in the vaccine group (n=5) compared to the placebo group (n=1) within 7 days of vaccination is possibly related to the vaccine. Numerical imbalances were reported between vaccine and placebo recipients for thromboembolic events (15 versus 10) and tinnitus (6 versus 0).

Based on currently available information, a contributory effect of the vaccine could not be excluded, although the imbalance was small (representing a difference of 0.06% of vaccine recipients vs. 0.05% of placebo recipients), and many of the participants had predisposing conditions. FDA will recommend surveillance for further evaluation of thromboembolic events with deployment of the vaccine into larger populations. There were no other notable patterns or numerical imbalances between treatment groups for specific categories of adverse events that would suggest a causal relationship to Ad26.COV2.S. As of February 5, 2021, a total of 25 deaths were reported in the study (5 vaccines, 20 placeboes). These deaths represent events and rates that occur in the general population of individuals in these age groups and include 7 deaths in the placebo group due to COVID-19 infection. Nonfatal serious adverse events, excluding those due to COVID-19, were infrequent and balanced between treatment groups with respect to rates and types of events (0.4% in both groups). A serious event of a hypersensitivity reaction, not classified as anaphylaxis, beginning 2 days following vaccination was likely related to receipt of the vaccine.

(FDA Briefing Document Janssen Ad26.COV2.S Vaccine for the Prevention of COVID-19 2021)

Based on the available data from randomized controlled trials that were double-blinded, we can conclude without bias that each vaccine trial demonstrated that their vaccines are safe and efficacious. "When drugs or vaccines are being trialed for their effectiveness, there are typically several stages. Double-blind trials are seen as the most reliable type of study because they involve neither the participant nor the doctor knowing who has received what treatment. The aim of this is to minimize the placebo effect and minimize bias. In double-blind trials, the treatment patients have is unknown to both patients and doctors until after the study is concluded. This differs from other types of trials, such as simple blind trials where only the patients are unaware of the treatment they are receiving, whereas the doctors know. Double-blind trials are a form of randomized trials and can be 'upgraded' to triple-blind trials, in which the statisticians or data clean-up personnel are also blind to treatments. To be effective, it is generally recommended that double-blind trials include around 100-300 people. If treatments are highly effective, smaller numbers can be used but if only 30 or so patients are enrolled, the study is unlikely to be beneficial. The assignment of patients into treatments is typically done by computers, where the computer assigns each patient a code number and treatment group. The doctor and patients only know the code number to avoid bias, hence allowing the study to be double-blind. Double-blind trials can come in different varieties. Double-blind, placebo-controlled studies involve no one knowing the treatment assignments to remove the chance of placebo effects. In a double-blind comparative trial, a new treatment is often compared to the standard drug. This allows researchers to compare an established drug to a new one to establish which one is more advantageous. However, unlike double-blind, placebo-controlled trials, they are not very good at statistically evaluating if a treatment is effective overall." (Ryding, B.Sc. 2021)

Science's job is to falsify data, and that requires scrutinizing that data over and repeatedly. If the data is unfalsifiable, then it's accepted, but the one moment it can't be repeatedly tested, it is no longer accepted. Double-Blinded Randomized Control Trials element's chance and deals with facts based on honest conclusions, not speculations or assumptions. It's not driven by opinions or appeals to emotions. If I'm looking to verify scientifically if mRNA vaccines are safe for my family and friends, I want to know if example a has been tested if so, where is the data that supports what is being said, and example b has met every safety measure required by not only the FDA. Having this data during the middle of a pandemic made it necessary for me to verify a solution to a much bigger problem. The benefits of receiving my mRNA vaccine outweighed the risk of being infected by Covid-19 and potentially having underlining conditions, facing hospitalization, the threat of death, and possibly infected relatives who may also face the same consequences. Besides, herd immunity cannot be achieved without a vaccine, and mRNA vaccines are and has proven its worth as 3rd generation vaccine technology will lead us into the future better prepared than we were in the early to mid to late 1700s or even just a year ago. Another benefit of mRNA vaccines is the fact they can be made faster and provide an instant real-time solution to further potential deadly pathogens. That is a huge benefit, and as the population increases, we will unearth or discover new viruses that will threaten our survival, and this technology helps resolve that issue.

Chapter V
Review

1). Do clinical trial results show whether vaccines are effective?

2). Do mRNA vaccine alter our DNA?

3). What are Randomized Trials?

4). What are 1st Generation Vaccines?

5). What are 2nd Generation Vaccines?

6). What are 3rd Generation Vaccines?

7). What was the name of the Dr. Who traveled from Hungary to work on mRNA?

8). What was the name of the person Cotton Mathers asked if he was familiar with smallpox?

9). Did Africans have knowledge of Inoculation?

10). Did Moderna, Pfizer, and Johnson & Johnson provide safety data from their randomized double blinded trials?

11). True or False: Attenuated pathogens, full organisms or inactivated bacterial toxin, which are effectively immunogenic, are used in making these vaccines.

12). True or False: Future mRNA vaccine technology may allow for one vaccine to provide protection for multiple diseases, thus decreasing the number of shots needed for protection against common vaccine-preventable diseases.

Chapter V Sources

- Ryding, Sara. "What Is a Double-Blind Trial?" News, 19 Mar. 2021, www.news-medical.net/health/What-is-a-Double-Blind-Trial.aspx.

- Flint, S. Jane, et al. Principles of Virology. Fifth ed., I, American Society for Microbiology, 2020.

-

- Tahamtan, Alireza, et al. "Home." Journal of Archives in Military Medicine, Kowsar, 26 Sept. 2017, sites.kowsarpub.com/jamm/articles/12315.html.

-

- Gray, Gregory, and Dean Erdman. "Adenovirus Vaccine." Adenovirus Vaccine - an Overview | ScienceDirect Topics, 2018, www.sciencedirect.com/topics/neuroscience/adenovirus-vaccine#:~:text=The%20adenovirus%20vaccine%20strains%20are,to%20over%2010%20million%20people.

- Dutta, Dr. Sanchari Sinha. "What Are Adenovirus-Based Vaccines?" News, 10 Mar. 2021, www.news-medical.net/health/What-are-Adenovirus-Based-Vaccines.aspx.

- Hewings-Martin, Ph.D., Yella. "How Do MRNA Vaccines Work?" Medical News Today, MediLexicon International, 18 Dec. 2018, www.medicalnewstoday.com/articles/how-do-mrna-vaccines-work#Addressing-stability-and-safety.

- Herbert, Eugenia W. "Smallpox Inoculation in Africa: The Journal of African History." Cambridge Core, Cambridge University Press, 22 Jan. 2009, www.cambridge.org/core/journals/journal-of-african-history/article/abs/smallpox-inoculation-in-africa/E43D8B3146D1EC4649699AD758E3B37A.

- Division of Viral Diseases, National Center for Immunization and Respiratory Diseases (NCIRD). "Understanding MRNA COVID-19 Vaccines." Centers for Disease Control and Prevention, Centers for Disease Control and Prevention, 4 Mar. 2021, www.cdc.gov/coronavirus/2019-ncov/vaccines/different-vaccines/mrna.html#:~:text=Future%20mRNA%20vaccine%20technology%20may,to%20target%20specific%20cancer%20cells.

- Boyle, Patrick. "MRNA Technology Promises to Revolutionize Future Vaccines and Treatments for Cancer, Infectious Diseases." AAMC, 29 Mar. 2021, www.aamc.org/news-insights/mrna-technology-promises-revolutionize-future-vaccines-and-treatments-cancer-infectious-diseases.

- Komaroff, Anthony. "Why Are MRNA Vaccines so Exciting?" Harvard Health, 11 Dec. 2020, www.health.harvard.edu/blog/why-are-mrna-vaccines-so-exciting-2020121021599.

-

- Pardi, Norbert, et al. "MRNA Vaccines - a New Era in Vaccinology." Nature News, Nature Publishing Group, 12 Jan. 2018, www.nature.com/articles/nrd.2017.243.

- Nahm, Moon. "COVID-19 MRNA Vaccines: How Could Anything Developed This Quickly Be Safe? - News." UAB News, The University of Alabama at Birmingham, 25 May 2021, www.uab.edu/news/youcanuse/item/12059-covid-19-mrna-vaccines-how-could-anything-developed-this-quickly-be-safe.

- Garde, Damian, and Jonathan Saltzman — Boston Globe. "The Story of MRNA: From a Loose Idea to a Tool That May Help Curb Covid." STAT, 7 Jan. 2021, www.statnews.com/2020/11/10/the-story-of-mrna-how-a-once-dismissed-idea-became-a-leading-technology-in-the-covid-vaccine-race/.

- Quality Promotion (DHQP), Centers for Disease Control and Prevention, National Center for Emerging and Zoonotic Infectious Diseases (NCEZID), Division of Healthcare. "U.S. Vaccine Safety - Overview, History, and How It Works." Centers for Disease Control and Prevention, Centers for Disease Control and Prevention, 9 Sept. 2020, www.cdc.gov/vaccinesafety/ensuringsafety/history/index.html.

- *Tx Inc., Moderna. Vaccines and Related Biological Products Advisory Committee Meeting , vol. 1, 17 Dec. 2020, pp. 1–54.*

- BioNtech, Pfizer. Vaccines and Related Biological Products Advisory Committee Meeting, Pfizer and BioNTech, 10 Dec. 2020, www.fda.gov/media/144245/download.

- Biotech Inc, Janssen. "Vaccines and Related Biological Products Advisory Committee Meeting." FDA Briefing Document Janssen Ad26.COV2.S Vaccine for the Prevention of COVID-19, Janssen Biotech, Inc., 26 Feb. 2021, www.fda.gov/media/146217/download.

- Flint, S. Jane, et al. *Principles of Virology*. Fifth ed., II, Wiley/ASM Press, 2020.

Myths & Pseudoisms Related to Covid-19
Chapter VI

Over the past few years an online campaign of misinformation has dominated timelines regarding vaccines. Everything began to heighten once the Brooklyn and Samoa measles outbreak began to dominate the headlines. Del Bigtree and Robert Kennedy took a crew of anti-vaxxers on a tour to fight against vaccine requirements in the New York area. Their battle or plight was more so about privilege than health and this was evident as their social media campaigns made a mess of misrepresenting vaccines and vaccine safety. To be honest scientist have stayed out of these conversations for years not wasting time entertaining people who make outlandish claims, but in the middle of the pandemic something changed and the narrative continues to be addressed.

An article written by Center for Countering Digital Hate called The Disinformation Dozen opted for a ban of 12 misinformation pushing people who tended to play on the ignorance of others. On social media anti-vaxx followers has reached more than 50 million people and mostly targeting the black community. However, the 'Dirty Dozen' were responsible for over 65% of the misinformation spread on social media. When the emergence of Sars-CoV2 Covid-19 hit it became evident that the noise would only get louder and louder from that direction. Anti-vaccine talking points were all similar and narrowed down to safety, ingredients, privilege, and with mRNA vaccines specifically the process.

From late 2019 until recently I have partaken in a series of teaching moments what some call debates on the subject matter. I've been grateful to sit on panels with Medical Doctors (Dr. Bonita Coe Clevland, Oh, Dr. Leleka Doonquah Washington D.C. & Dr. Lynn Milliner Cleveland OH) who have been on the front lines since day 1 fighting Covid-19. Brother Kevin Chill Heard arranged those conversations on his platform which is called ChillTalk on Facebook and Youtube. Brother Kevin Chill Heard continued to allow us to present a sound argument to help spread the truth about vaccines. In several discussion with a group of brothers and a sister who argued for a natural solution we realized that we were dealing with regurgitated information pushed by the 'Dirty Dozen'.

We spent our entire 2020 year trying to inform the misinformed who parroted misinformation around as if they were paid to do so. I've heard almost all of the silliest claims made by people who believe what they hear without fact checking the source and Covid-19 exposed the ignorance that existed among us. Brother Ankh and myself also had a discussion with 2 Pharmacologist on 360 info with a Louis Jefferson and Dr. Metasebya Solomon regarding vaccine science. In that discussion it was revealed to us that neither of them were aware of the mRNA pre-clinical trial data (claiming steps have been skipped), that the vaccine was not a vaccine (would modify your DNA), and Louis Jefferson attempted to claim that mRNA vaccines have the potential to cause Ribosome misreads which would lead to cancerous reactions within the body. In this chapter we will deal with claims made by both Louis Jefferson and Dr. Metasebya Solomon and a host of other anti-vaccine claims regarding Sars-Cov2 (origins), 5G spreading Covid-19, mRNA vaccines, nurses dying, and much more.

"The CCDH (Center for Countering Digital Hate) is a not-for-profit NGO that seeks to disrupt the architecture of online hate and misinformation. Digital technology has changed forever the way we communicate, build relationships, share knowledge, set social standards, and negotiate and assert our society's values. Digital spaces have been colonized and their unique dynamics exploited by fringe movements that instrumentalize hate and misinformation. These movements are opportunistic, agile and confident in exerting influence and persuading people. Over time these actors, advocating diverse causes - from anti-feminism to ethnic nationalism to denial of scientific consensus - have formed a Digital Counter Enlightenment. Their trolling, disinformation and skilled advocacy of their causes has re-socialized the offline world for the worse. The Center's work combines both analysis and active disruption of these networks. CCDH's solutions seek to increase the economic, political and social costs of all parts of the infrastructure - the actors, systems and culture - that support, and often profit from hate and misinformation." (CCDH 2020)

This group focuses on misinformation spread by anti-vaxxers on all social media platforms. After compiling enough data to dwindle down the number of influencers on involved the CCDH has concluded on 12 individuals responsible for more than 65% of the disinformation found online. More specifically, 73% of Facebook post, 17% of tweets, and re-shares all over instagram. "Our analysis of over 812,000 posts extracted from Facebook and Twitter between 1 February and 16 March 2021 shows that 65 percent of anti-vaccine content is attributable to the Disinformation Dozen. This shows that while many people might spread anti-vaccine content on social media platforms, the content they share often comes from a much more limited range of sources. Exposure to even a small amount of online vaccine misinformation has been shown by the Vaccine Confidence Project to reduce the number of people willing to take a Covid vaccine by up to 8.8 percent." (CCDH 2020)

"At the outset of this research, we identified a dozen individuals who appeared to be extremely influential creators of digital anti-vaccine content. These individuals were selected either because they run anti-vaccine social media accounts with large numbers of followers, because they produce high volumes of anti-vaccine content or because their growth was accelerating rapidly at the outset of our research in February. 1. Joseph Mercola 2. Robert F. Kennedy, Jr. 3. Ty and Charlene Bollinger 4. Sherri Tenpenny 5. Rizza Islam 6. Rashid Buttar 7. Erin Elizabeth 8. Sayer Ji 9. Kelly Brogan 10. Christiane Northrup 11. Ben Tapper 12. Kevin Jenkins.

Out of the 12 people involved in misleading the masses I was only unfamiliar with one person. That person is Kevin Jenkins and he is like Rizza Islam, Kevin has partnered up with Robert Kennedy and the boys to drive the dis-information campaign targeting African Americans. After reviewing some of his claims I began to understand where this claim that: Hank Aaron was being paid to indorse the Covid-19 vaccines began. Even calling the Ancestor a modern day slave catcher. Kevin like his predecessors only provide rhetoric with absolutely zero supportive evidence.

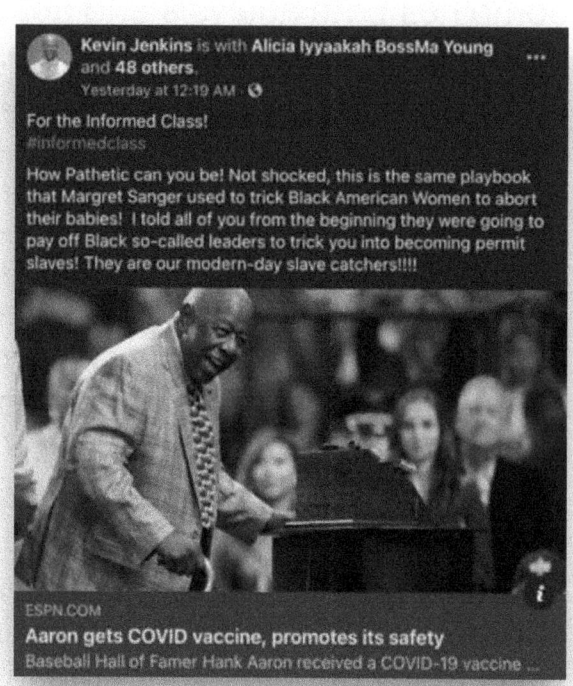

Speaking of Hank Aaron, a myth following his untimely death began to circulate social media immediately claiming that he died after receiving Moderna m1273 vaccine. "One person wrote on Facebook: "Our Brother Hank Aaron – RIP – wanted to be an example and an inspiration to Black People by taking the COVID-19 vaccine. Unfortunately, he may have become a clear example to Black People of why this vaccine CANNOT be trusted. … Two weeks after receiving his first Moderna shot, Hank Aaron died in his sleep. No cause of death is given, but you do the math." Another Facebook user, who juxtaposed one news article about Aaron getting the vaccine and another about his subsequent death, wrote: "I hope people receive the message that this mans death should send to the rest of us!" And Robert F. Kennedy Jr., a vaccine skeptic who has promoted COVID-19 misinformation, tweeted that "#HankAaron's tragic death is part of a wave of suspicious deaths among elderly closely following administration of #COVID #vaccines."

Kennedy's tweet linked to a post by his organization, the Children's Health Defense, baselessly suggesting Aaron may have died as a result of the vaccine. While the exact cause of Aaron's death has not been determined, health officials do not believe it was related to him being vaccinated. An official with the Fulton County Medical Examiner's office, which examined Aaron's body following his death, told FactCheck.org by phone that he died "due to natural causes." In a statement emailed to AFP Fact Check, the county medical examiner, Karen Sullivan, said: "There was no information suggestive of an allergic or anaphylactic reaction to any substance which might be attributable to recent vaccine distribution." She added: "In addition, examination of Mr. Aaron's body did not suggest his death was due to any event other than that associated with his medical history." WSB-TV in Atlanta reported that Aaron had a history of prostate issues and hypertension. In addition, he received treatment for osteoarthritis of the knee and used a wheelchair later in life. The Morehouse School of Medicine, which is affiliated with the Atlanta clinic where Aaron was immunized, also said the vaccine was not a factor in his death. The historically Black medical school emailed us a statement that said: "Mr. Aaron was a public health advocate and worked with us to help bridge the health equity gap in Atlanta and around the world. His passing was not related to the vaccine, nor did he experience any side effects from the immunization. He passed away peacefully in his sleep." Aaron was among more than 20 civil and human rights activists — all over age 75 — who were vaccinated at the new health facility to "promote vaccine acceptance in the Black community," the medical school said in a Jan. 5 press release." (Gore 2021)

Immediately after his death and I saw a post stating such I questioned the validity of the claim. What autopsy was done so soon that one can make that assessment. The family had not put out an official statement nor blamed the vaccine and here we have it people already disrespecting the Ancestor right after his sudden transition.

Dr. Ben Tapper a chiropractor from Omaha has been very loud on social media. His field is not in virology, epidemiology, or immunology. Like many others who make claims people see the initials D and R and assume that the validity of their claims are qualified despite the field of study they are a professional in. One of his most famous claims that circulated the internet was about PCR test. In a video circulating all over social media he introduces himself as a Doctor of PCR testing. "Speaking to a city council in the USA and saying that he is a medical doctor, Ben Tapper says in his video that spread widely on social media, the PCR test gives false positive results and should not be used in the detection of diseases. Making a dedication to Kary Mullis, the inventor of the test, Tapper also claims that the vaccines are forcibly sold to the states by companies. Tapper makes a series of claims that will feed science and pandemic skepticism throughout his speech loaded with sensational statements. First of all, Ben Tapper is not a medical doctor. He studied business administration and later started his father's profession as a chiropractor . Chiropractic spine, bones and muscles by applying the press suggesting that regulate the nervous system, an alternative medicine method and "pseudo-science" is called . It is not necessary to study medicine to be able to practice this profession, it is enough

to get an education from an institution providing chiropractic education. There is an institution in Tapper's resume that states that he received this training , but the doctorate degree obtained here has no academic equivalent. In other words, Ben Tapper lacks the academic background and expertise to support his claims. Tapper's first claim is that PCR tests, which are widely used in the diagnosis of coronavirus, give false positive results and that the pandemic is therefore not real but delusional. However, as Teyit
has examined before , the false positive rate of PCR tests is very low. The most reliable method to detect Covid-19 is again the PCR test. This method, which allows to detect the genetic material of the pathogen causing the disease, is a reliable way to understand whether there is a virus at the sample point . The margin for error is low when done correctly. Although it may give a false negative result when there is not enough virus in the sample area, the probability of the opposite is quite low. Another claim by Tapper is against Kary Mullis, the inventor of the PCR test. Tapper claims that Mullis said that PCR tests should not be used for diagnosis because they can detect any disease in anyone. Another claim of Ben Tapper is about vaccine companies. Tapper, who is also against masks, claims that Moderna and Pfizer companies are forcing states to make vaccination mandatory, just like masks. First of all, masks are not mandatory in every state in the USA. Each state has its own regulations. The states themselves decide whether to get the vaccine or from which company. Many countries
have already purchased large quantities of vaccines voluntarily, through preliminary agreements with vaccine companies . Making the vaccine compulsory in the community also depends on what states and countries can and cannot do under their laws .On Tapper's social media accounts , it can be seen that he regularly posts anti-vaccine and anti-mask posts and shares conspiracy theories." (Zubeyir 2020)

Christian Northrup is an obstetrics and gynecologist who made claims in a recent video that circulated Facebook and Instagram just as quickly as any other claim made by the Dirty Dozen. She is popular for making the 5G and Covid-19 claim. In her video she made numerous claims about Covid-19 vaccines that simply was not true.

"The video is a 4-minute clip of the 37-minute long interview with Dr. Christiane Northrup, an obstetrician- gynaecologist who has earlier featured as a health expert on "The Oprah Show". She claimed that the upcoming COVID-19 vaccine will alter the human DNA. The vaccine henceforth, is set to control our movements has gone viral on social media. Claim 1: The RNA vaccine will fundamentally change people's DNA. Fact: According to experts, mRNA vaccines do not alter human DNA. Such misinformation is regularly aired on social media to sow doubts/mistrust even before the vaccine is available to the public. Thus, the claim made in the post is FALSE. Claim 2: The toxic metals used in the vaccine will create antennas in the human body that will be detected by 5G technology. Fact: Multiple conspiracies and speculations have been doing rounds on social media but in fact, 5G technology has nothing to do with COVID-19. Thus, the claim made in the video is FALSE. Claim 3: The non-human DNA will turn humans into chimers. Fact: There is no scientific evidence that the mRNA vaccine will introduce non-human/animal DNA into human bodies. Thus, the claim made in the video is FALSE. Claim 4: Vaccine to contain nano particles which in turn act as antennas will collect biometric data that will further be traded for cryptocurrency. Fact: While there is a patent application that mentions technology allowing to monitor & track people's activity in exchange for cryptocurrency, there is no mention of nano particles or implanted chips in the body. Thus, the claim made in the video is FALSE. Claim 5: The patented dye Luciferase produced by MIT will help countries to check on who has been vaccinated. Fact: In 2019, Massachusetts Institute of Technology did develop a novel way to record patient's MMR vaccine history using the luciferase enzyme. However, the same has not been approved for COVID-19 vaccine. Thus, making the claim MISLEADING." (Kalidoss 2020)

Kelly Brogan is an author and a promoter of alternative medicines. She is also known as the person that denies reality, however she is responsible for continuing the claim that Covid-19 vaccines make women infertile. Originally this claim was falsely made by a German epidemiologist. "Last December, a German epidemiologist said the COVID-19 vaccines might make women's bodies reject a protein that's connected to placenta, therefore making women infertile. He thought this because the genetic code of the placenta protein, called syncytin-1, shares a hint of similarity with the genetic code of the spike protein in COVID-19. If the vaccines caused our bodies to make antibodies to protect us from COVID-19, he thought, they could also make antibodies to reject the placenta. This, however, was a theoretical risk that was completely disproven in the clinical trials and continues to be disproven in real time as more women of child-bearing age become fully vaccinated. "It's inaccurate to say that COVID-19's spike protein and this placenta protein share a similar genetic code," says D'Angela Pitts, M.D., a maternal fetal medicine specialist with Henry Ford Health System. "The proteins are not similar enough to cause placenta to not attach to an embryo." "Women who participated in the COVID-19 clinical trials were able to conceive after vaccination," says Dr. Pitts. "We also have many patients here at Henry Ford who got vaccinated and then became pregnant afterwards. Some are in their first trimester, some are now in their second trimester. There's no evidence to show that the COVID-19 vaccines lead to reduced fertility." The mechanism used to create the Pfizer and Moderna vaccines—called mRNA technology—is not new. It has been used widely for decades in different treatments, so there is plenty of data on the use of mRNA technology and fertility. And with the COVID-19 vaccines, specifically, research shows there are no adverse outcomes or safety issues in connection with reproductive health. For years, women have routinely and safely gotten vaccinated before pregnancy and during pregnancy. The only type of vaccination doctors don't recommend getting while pregnant is live vaccine (which is a strain of a live, weakened virus) because it could affect the fetus. The COVID-19 vaccines do not contain live virus. "There is no evidence that shows getting one of the COVID-19 vaccines will cause infertility or even cause complications that would require fertility workup," says Dr. Pitts. "I recommend that young women, millennials and Gen Z'ers, get the vaccine. I've seen evidence of younger and older women getting the vaccine and having no problems conceiving. I myself am a millennial woman who wants children in the future, and I did not hesitate to get vaccinated—I was very excited to get it." (Henry Ford Health Staff 2021)

Sayer Ji is a holistic health researcher and also has one of the largest alternative medicine health websites online called Greenmedinfo.com. "GreenMedInfo, an alternative health website that has published articles claiming vaccines cause autism, published an article on December 6th warning about the adverse side effects of the coronavirus vaccines, including death. The article has been shared thousands of times by anti-vaccine Facebook pages and groups. We rate this article as mostly false and misleading. Although the vaccines were developed in record time, it doesn't mean they're unsafe. The FDA is still monitoring for potential adverse side effects from the vaccines in development. So far, it has not found that any of the COVID-19 vaccines seeking approval cause death. Ji also cited an FDA briefing document on the vaccine developed by Pfizer and its partner, BioNTech. The United Kingdom on Dec. 8 became the first country to start distributing the vaccine, and Canada approved it the next day. In the document, the FDA reported that six of 43,448 participants died during clinical trials of the vaccine held from April to November. Two had been given the vaccine, and four had received a placebo. Both participants who received the vaccine and died were older than 55; one died from cardiac arrest, the other from arteriosclerosis. The vaccine was not listed among the causes of their deaths, which the FDA said were characteristic of their age groups. Although the agency did note some minor side effects associated with the vaccine, including fatigue and headache, it said there were "no specific safety concerns identified that would preclude issuance" of emergency use authorization. The FDA's assessment supports Pfizer's Nov. 18 announcement that its vaccine was safe and 95% effective at preventing the coronavirus. The agency will meet Dec.

10 to discuss whether to approve the emergency-use authorization, after which it will be assessed by the Centers for Disease Control and Prevention's Advisory Committee on Immunization Practices and continue to be studied in clinical trials. "The FDA would only issue an EUA if the vaccine has demonstrated clear and compelling efficacy in a large well-designed phase 3 clinical trial," the FDA told PolitiFact. "If an EUA is issued, the process will not be rushed, and no shortcuts will be taken around having the relevant phase 3 efficacy results." GreenMedInfo was created in 2008 and has

published several articles that falsely claim there's a link between vaccines and autism. In 2018, Pinterest banned the site for violating its policies against anti-vaccine misinformation. The website's statement contains an element of truth but ignores critical facts that would give a different impression. We rate it Mostly False." (Jones 2020)

Erin Elizabeth another alternative medicine promoter who runs a Health Nut News site. "The claim: Official with Operation Warp Speed said 'All of America must receive the vaccine within 24 hours of distribution'. After coronavirus vaccine candidates from Pfizer and Moderna showed up to 95% efficacy, officials from Operation Warp Speed, the White House-led initiative to quickly develop COVID-19 vaccines and treatments, provided an update about their efforts to assist with vaccine production and distribution. Army Gen. Gustave F. Perna, the chief operating officer for Operation Warp Speed, provided extensive information about vaccine distribution plans at a panel on Nov. 18. Full footage of the update is available from the Department of Defense and C-SPAN. Erin Elizabeth – an author and public speaker is also known as "Health Nut News" –posted a one-minute clip of his remarks on Facebook and Instagram. "All of America must receive the vaccine within 24 hours of distribution – there is no have and have Nots," she wrote, quoting the video. Elizabeth added that she is "not a fan" of Operation Warp Speed, and "not too thrilled" about the idea of mandated coronavirus vaccines. Breana Janel also posted a short clip of the panel on Facebook, highlighting phrases that Perna uttered, including "All of America must receive." "Nope... don't like that," she wrote. The posts both emphasize the quote, "All of America must receive the vaccine within 24 hours of distribution." But Perna was not referring to mandated vaccination – or even receiving the vaccine on an individual level. He was talking about the importance of quickly distributing the vaccine to each of the nation's 64 jurisdictions – including the 50 states, eight territories and six major metropolitan areas. "Distribution will occur from our fill-finish site and will go to all 64 jurisdictions that we are responsible to ensure they get vaccine for within 24 hours. Everybody – every jurisdiction – will have access immediately," he explained. At the end of his remarks, Perna summarized his previous points and shared what he told his team are the "two things we must have." That's when he said the quote in the posts. "One, upon emergency use authorization, all of America must receive vaccine within 24 hours, as I said," he said. "I think this is incredibly important – fair and equitable distribution of the vaccine throughout the country simultaneously. Based on our research, the claim that an official with Operation Warp Speed said "All of America must receive the vaccine within 24 hours of distribution" is rated MISSING CONTEXT, because without additional context it is misleading. It's a bona fide quote from Army Gen. Gustave F. Perna – but his comments were in reference to the necessity of immediate vaccine access and "equitable distribution" across the nation, not mandatory vaccination." (Caldera 2020)

Rashid Buttar is an osteopathic physician who I'm very familiar with. Buttar has a unique situation because his ex-wife now, but years ago got their son vaccinated against his will. Post vaccination he claimed that the vaccines caused his sons autism that he later would cure via holistically. Buttar is known for making claims about Covid-19 vaccines and Dr. Fauci. "The interview with Dr. has been deleted from Youtube several times. Rašidu Butaru," Kārlis Brants invites Facebook users to watch this video. The video, which the Latvian version has shared with almost half a thousand people, promises to reveal the hidden truth of the government and the media about the pandemic caused by Covid-19. It contains a number of false and unprovable allegations, including that Dr. Anthony Fauch is responsible for causing the pandemic. Butar, a guest on the show who is listed as "one of the top 50 doctors in the United States," is an osteopath who has twice been reprimanded by the North Carolina

Medical Council for his unethical treatment of patients. The US Food and Drug Administration (FDA) has warned him about the unauthorized promotion of unapproved drugs. However, due to his alternative treatment methods, he has repeatedly come to the attention of the media. He uses an unproven method of treatment for almost every medical diagnosis, from autism to cancer. This is called 'chelation therapy', which removes metals from the body. During the Covid-19 pandemic, this is not the first time he has distributed a conspiracy video that Youtube deleted because of its lying."Fauci is directly responsible not only for the pandemic, but also for the response."Not true - the statement is not true, there is no evidence, the author of the statement lies or unknowingly misleads Buttar begins by insulting that Anthony Fauci, director of the U.S. National Institute of Allergy and Infectious Diseases Hospitals, is to blame for the pandemic because "already in 2015, he illegally allowed funding for research that has now led to Covid-19. "The Fauci-led institute does study the spread of viruses.

However, a study examining whether coronaviruses can spread from animals to humans is just one of a number of studies funded by the US Institute of Health, with at least 20 studies published in the last six years. In addition, they were funded following the worldwide spread of another type of coronavirus, SARS, in 2003. The idea that the virus was artificially created in the laboratory, as Buttar claims, has already been refuted by. Buttar goes on to say that in his 2017 speech , Fauci predicted that the current US presidential administration would face a pandemic. Buttar describes it as a suspicious coincidence, indicating that the spread of the virus had been planned in advance. In a speech at Georgetown University, Fauci did warn that the US government might face an outbreak of an unknown disease in the coming years, but, like other speakers, called for awareness of the potential risks and for them to be prepared in time. Fauci is not the only one who has spoken on this topic before.

Several scientists have warned about the possibility of a global pandemic. "There is no virus that jumps 6 or 12 feet away!" Rather, it is not true - the statement contains cryptic truths, but does not take into account relevant facts and / or context, so the statement is

misleading or out of context. In this way, Buttar tries to justify that social distancing is unjustified. 6 feet or 1.8 meters is not a magical distance that guarantees complete protection, but
scientists justify that most of the droplets that a person emits when coughing or sneezing land no further than two meters. However, other studies show that squeegee drops can spread even more than two meters. "The PCR test cannot be used for diagnostic purposes." Not true - the statement is not true, there is no evidence, the author of the statement lies or unknowingly misleads. The interviewee states that PCR tests, which are also used in Latvia to detect the presence of the virus, cannot be used for diagnostic purposes, but only in
cases when "a particle has already been identified with a genome sequence". The World Health Organization has recognized PCR tests as the most reliable method for detecting the presence of this virus. "No one has reported even one death directly from the virus." Not true - the statement is not true, there is no evidence, the author of the statement lies or unknowingly misleads. The fact that the death is due to Covid-19 , is justified both by the World
Health Organization (WHO) and the European Center for Disease Prevention and Control Center (ECDC) guidelines and are guided by each country on its own established procedures. According to WHO and ECDC guidelines, people with a laboratory-confirmed positive result for SARS-CoV-2 coronavirus and Covid-19 as the primary or secondary cause of death have been identified as victims of the virus . On the other hand, the assertion that 'doctors and nurses are required to correct death certificates so that the primary cause of death is Covid-19 ' is unfounded. Even US President Donald Trump, whose reasoning Butar calls for in the same video not to be questioned and listened to, has said that the number
of Covid-19 victims is accurate. In April, the US Center for Disease Control (CDC) published new guidelines to be registered with the coronavirus related death: "In cases, if you can not determine the exact Covid-19 diagnosis, but it is for the suspect, the death certificate is permissible to write Covid-19 as a" possible "or "Probable" cause of death." "If you have received the flu vaccine, especially the triple flu vaccine, it is very likely that you will get a false-positive result in the CV-19 test." Not true - the statement is not

true, there is no evidence, the author of the statement lies or unknowingly misleads. A similar claim by Re: Check has already been refuted in this article. There are insufficient scientific data to support a causal relationship between the influenza vaccine and Covid-19. A study published in January to see if the flu vaccine increases the risk of getting another respiratory virus failed to prove a correlation." (Vebere & Baltica 2020)

Rizza Islam heavily influenced by Minister Farrakhan and directly aligned with Kennedy has been one of the most influential disinformation campaigners on social media. He has used platforms all over to spread misinformation about vaccines, vaccine science, and more. People have reposted clip after clip after clip of Rizza making the craziest claims. But he has continuously rehashed the Tuskegee Experiment in the black community. He parrots the NOI's claims in the community that range from calling the vaccine an experiment and even 5g. He does a great job at recirculating Robert Kennedy and Del Bigtrees talking points so much so its garnered him national attention. Rizza has claimed that vaccines are used to depopulate the world without any scientific evidence to support his and the Nations claims. He made his name known in a video where he talked about a slew of vaccine claims in basically one paragraph.

FULL CLAIM: "According to the 1986 National Childhood Vaccine Injury Act, you cannot sue a vaccine manufacturer if the vaccine has harmed and/or killed someone"; "An infertility drug [in the tetanus vaccine] sterilized over 500,000 Kenyan women under the Bill and Melinda Gates Foundation"; "Vaccines are 'unavoidably unsafe." - "Claim 1 (Inaccurate): According to the 1986 National Childhood Vaccine Injury Act, you cannot sue a vaccine manufacturer if the vaccine has harmed and/or killed someone." The video alleges that the National Childhood Vaccine Injury Act prevents people from suing a vaccine manufacturer for vaccine injuries. This is false, as the Act still permits individuals to pursue legal action against a vaccine manufacturer under certain conditions, for example, if the individuals reject the decision made by the vaccine court, or if a vaccine manufacturer has been shown to be negligent. Furthermore, vaccines are not associated with autism or a higher risk of cancer, contrary to the claims in the video. Claim 2 (Inaccurate): "The VAERS system […] established the vaccine court" The claim is inaccurate. Firstly, VAERS, or the Vaccine Adverse Events Reporting System in full, allows people to report adverse events that occur following vaccination. It is not a legal entity. The VAERS website explains: "The Vaccine Adverse Events Reporting System (VAERS) is a national early warning system to detect possible safety problems in U.S.-licensed vaccines. VAERS is co-managed by the Centers for Disease Control and Prevention (CDC) and the U.S. Food and Drug Administration (FDA). VAERS accepts and analyzes reports of adverse events (possible side effects) after a person has received a vaccination. Anyone can report an adverse event to VAERS." Claim 3 (Inaccurate): "The CDC admitted that over 90 million people that received a polio vaccine came down with a cancer type of circumstance due to a cancer-causing agent known as SV40, also known as simian virus 40 in the polio vaccine." This claim is misleading and was previously reviewed by Health Feedback here. As explained in our review, certain batches of polio vaccine were indeed contaminated with simian virus 40, leading millions of Americans to become exposed to SV40. However, "there is no reliable evidence that SV40 causes cancer in humans," said Michael Imperiale, a professor of microbiology and immunology at the University of Michigan. Dana-Farber Cancer Institute, a cancer treatment and research institute, stated that "vaccines won't give you cancer." Claim 4 (Inaccurate): "MMR vaccine increased autism rate by 236% in African-American boys compared to Caucasian boys" This claim was also previously reviewed by Health Feedback and found to be inaccurate. It was also fact-checked by Snopes and found to be inaccurate. It is based on a study by chemical engineer Brian Hooker, that was later retracted by the journal. The journal editors explained their decision to retract the study, stating, "There were undeclared competing interests on the part of the author which compromised the peer review process. Furthermore, post-publication peer review raised concerns about the validity of the methods and statistical analysis, therefore the Editors no longer have confidence in the soundness of the findings." The methodological flaws were covered in detail in this ScienceBlogs article: "Hooker analyzed a dataset collected to be analyzed by a case-control method using a cohort design. Then he did multiple subset analyses, which, of course, are prone to false positives. As we also say, if you slice and dice the evidence more and more finely, eventually you will find apparent correlations that might or might not be real. In this case, I doubt Hooker's correlation is real." "And, of course, there's no biologically plausible reason why there would be an effect observed in African-Americans but no other race and, more specifically than that, in African-American males. In the discussion, Hooker does a bunch of handwaving about lower vitamin D levels and the like in African American boys, but there really isn't a biologically plausible mechanism to account for his observation, suggesting that it's probably spurious." Claim 5 (Inaccurate): "An infertility drug [in the tetanus vaccine] sterilized over 500,000 Kenyan women under the Bill and Melinda Gates Foundation" This is another vaccine myth that has been debunked many times over. In fact, this article by Africa Check reports that the claim "is more than 20 years old and has been repeatedly debunked by [the] World Health Organisation (WHO) and others ever since." According to Africa Check, the claim is based on "a misunderstanding of a scientific study in India in 1994 that tested a birth control treatment." This study tested the efficacy of a birth control treatment using the hormone human chorionic gonadotropin (hCG), which is produced in the body at high levels during pregnancy and is also excreted in the urine. The detection of hCG in urine is the basis of home pregnancy test kits: "For the Indian trial, researchers used a protein similar to the tetanus toxin as a carrier for the hCG. This would then cause the woman's immune system to eliminate hCG to prevent pregnancy. The process was reversible, though. An American anti-abortion organization, claiming support from the Vatican, used this information to call for a congressional investigation into Mexico's tetanus vaccination programme. Human Life International claimed that the tetanus vaccine being administered contained hCG which would leave women infertile." In 2014, this claim was revived by some Kenyan bishops and a Catholic medical organization, who alleged that the tetanus vaccine contained hCG.

However, this claim was based on test results which turned out to be unreliable, because the method used was only appropriate for human samples, not vaccine samples. Claim 6 (Misleading): "Vaccines are 'unavoidably unsafe'" This is another common talking point among the anti-vaccine community, which interprets the term to mean that vaccines are dangerous. However, the term "unavoidably unsafe" is a legal term that has a specific meaning that differs from the anti-vaccine interpretation, and takes into account the risk/benefit tradeoffs, as we will see below. The term "unavoidably unsafe" comes from a comment to Section 402A of the Restatement (Second) of the Law of Torts, published by the American Law Institute: "Unavoidably unsafe products: There are some products which, in the present state of human knowledge, are quite incapable of being made safe for their intended and ordinary use. These are especially common in the field of drugs. An outstanding example is the vaccine for the Pasteur treatment of rabies, which not uncommonly leads to very serious and damaging consequences when it is injected. Since the disease itself invariably leads to a dreadful death, both the marketing and the use of the vaccine are fully justified, notwithstanding the unavoidable high degree of risk which they involve. Such a product, properly prepared, and accompanied by proper directions and warning, is not defective, nor is it unreasonably dangerous." Dorit Reiss, a professor of law at the University of California Hastings College of the Law, explained the implications of the comment in this 2013 post: "The last sentence is the important one: A vaccine whose benefits outweigh its risks is not unreasonably dangerous or defective – even if the risks are as frightening as those attributed to the Pasteur vaccine, let alone modern vaccines, with their much lower risks. […] Saying a product is 'unavoidably unsafe' makes it sound like the product is a bad one, when what the drafters meant was precisely the opposite: the comment was meant to apply only to ethical drugs or vaccines, i.e. where the benefits outweigh the risks." (Teoh 2020)

Rizza latest CLAIM: "The Johnson & Johnson vaccine has aborted fetal tissue from aborted babies from the retina as well as from kidney cells. Factually inaccurate: There aren't any cells or tissues from aborted fetuses in the Johnson & Johnson COVID-19 vaccine, according to the ingredient list. A cell line derived in 1985 from an aborted fetus is used during production, but the vaccine components are extensively purified during the process." (Rougerie 2021)

Sherri Tenpenny another osteopathic physician and anti-vaxxer claimed that the mRNA vaccines contained things in them that make people magnetized. She has also claimed that mRNA vaccines affect mens sperm. Like he counterparts she stays within the realm of conspiracy theories never providing evidence for her claims. Her claims have lead to some people placing a magnet on their arms where they received the vaccine. "Sherri Tenpenny: The COVID-19 vaccines make people "magnetized. They can put a key on their forehead, it sticks. They can put spoons and forks all over them, and they can stick."PolitiFact's ruling: False Here's why: An anti-vaccine activist falsely claimed during a hearing with Ohio state legislators that the COVID-19 vaccines are magnetizing people who get them. Dr. Sherri Tenpenny, an Ohio-based osteopathic physician who wrote a book called "Saying No to Vaccines," has been identified by the news site rating service Newsguard as a "super-spreader" of COVID-19 vaccine misinformation. A watchdog group at McGill University in Montreal found that she is one of 12 influencers responsible for 65% of anti-vaccine misinformation spread on Facebook, Instagram and Twitter. Tenpenny has previously pushed false claims that the COVID-19 vaccines can cause death and autoimmune disease, disrupt pregnancies and "shed" to affect unvaccinated people. Her latest comments came as she testified at the invitation of Ohio's Republican lawmakers in favor of a bill that would prevent businesses or the government from requiring proof of COVID-19 vaccination, according to the Columbus Dispatch. "I'm sure you've seen the pictures all over the internet of people who've had these shots, and now they're magnetized," Tenpenny said during the June 8 hearing. "They can put a key on their forehead, it sticks. They can put spoons and forks all over them, and they can stick.

Because now we think that there's a metal piece to that." Those claims are baseless. There are no metallic ingredients in any of the COVID-19 vaccines approved in the U.S. for emergency use, from Pfizer-BioNTech, Moderna and Johnson & Johnson. The Food and Drug Administration has published the ingredients for each online. "There's nothing there that a magnet can interact with," Thomas Hope, a vaccine researcher at Northwestern University, previously told AFP Fact Check. "It's protein and lipids, salts, water and chemicals that maintain the pH. That's basically it, so this is not possible." PolitiFact and several other fact-checkers previously debunked the videos and "pictures all over the internet" that Tenpenny cited as proof, which purported to show magnets sticking to vaccinated people. The social media posts about vaccine magnetism were so widespread that the U.S. Centers for Disease Control and Prevention addressed them on its website: "Can receiving a COVID-19 vaccine cause you to be magnetic? No. Receiving a COVID-19 vaccine will not make you magnetic, including at the site of vaccination which is usually your arm. COVID-19 vaccines do not contain ingredients that can produce an electromagnetic field at the site of your injection. All COVID-19 vaccines are free from metals such as iron, nickel, cobalt, lithium, and rare earth alloys, as well as any manufactured products such as microelectronics, electrodes, carbon nanotubes, and nanowire semiconductors. In addition, the typical dose for a COVID-19 vaccine is less than a milliliter, which is not enough to allow magnets to be attracted to your vaccination site even if the vaccine was filled with a magnetic metal." Florian Krammer, a professor of vaccinology at New York's Icahn School of Medicine at Mount Sinai, previously told PolitiFact the claims about vaccine magnetism were "utter nonsense." Other experts told AFP Fact Check that the metal objects highlighted in various online videos and images are likely sticking for other reasons. They could have tape or another adhesive on them, for example. Or they could appear to stick because of the oil on a person's skin. Later on in the Ohio House hearing, a nurse tried unsuccessfully to prove Tenpenny's theory by positioning a key and a bobby pin against her neck. "Explain to me why the key sticks to me.

It sticks to my neck too," she said, even as the key she pressed to her neck did not stick. Tenpenny also claimed that there is "some sort of an interface, 'yet to be defined' interface, between what's being injected in these shots and all of the 5G towers," and that the vaccines have caused thousands of deaths in the U.S. Both of those claims are inaccurate. Tenpenny did not immediately respond to a request for comment from PolitiFact. She told the Washington Post that she stood by her testimony. We rate her claim that the vaccines make people "magnetized" False." (McCarthy 2021)

Ty & Charlene Bollinger "The Bollingers are part of an ecosystem of for-profit companies, nonprofit groups, YouTube channels and other social media accounts that stoke fear and distrust of COVID-19 vaccines, resorting to what medical experts say is often misleading and false information. An investigation by The Associated Press has found that the couple work closely with others prominent in the anti-vaccine movement — including Robert F. Kennedy Jr. and his Children's Health Defense — to drive sales through affiliate marketing relationships.According to the Bollingers, there is big money involved. They have said that they have sold tens of millions of dollars of products through various ventures and paid out $12 million to affiliates. Tens of thousands of people ponied up cash for an earlier version of their vaccine video series, they said." (Smith & Reiss 2021)

FULL CLAIM: "The nature of this virus is unique in a way not possible through natural means"; "Gates was busily working to establish global digital ID infrastructure long before the outbreak, which is now being promoted as a way to combat the virus"; Bill Gates, who has previously funded harmful vaccines in developing countries, plans to use COVID-19 vaccines to surveil the population. Unsupported: The Bill and Melinda Gates Foundation is a founding member of the Digital Identity Alliance, which seeks to provide secure and private digital identification to citizens of the world to provide better access to services. The program does not include any tracking system, nor is it related to the vaccination programs funded by the Foundation. Inaccurate: The vaccination campaigns funded by Gates in Africa and India were safe. The severe events and deaths reported in the article are either false or were due to causes unrelated to the vaccines. (Carballo-Carbajal 2021)

Robert F. Kennedy Jr., is the leading anti-vaxxers in the world he has argued and debated every piece of scientific evidence without producing anything credible. Mentioned in the introduction a debate in Harlem held in Harlem in late 2019 I watched first hand how both Robert Kennedy and Del Bigtree rehash anti-vaccine arguments over and over while Samoa and Brooklyn had measles outbreaks.

"This article posted by The Truth About Cancer on 13 June 2020 is the third and last article in a series claiming that vaccines are dangerous and that individuals will be forced to accept a SARS-CoV-2 vaccine once developed. The article, titled COVID-19: Conspiracies, Vaccines, & Bill Gates, discusses a post by Old-Thinker News and an Instagram post from Robert F. Kennedy Jr. The many claims contained in the article have been mentioned in tenths of thousands of social media posts and videos, with some versions receiving more than 100,000 views.
The authors begin by falsely claiming that mainstream media and the U.S. government have acknowledged that the SARS-CoV-2 virus "probably originated in a secret bio lab in Wuhan". The basis for this statement is a 17 April editorial article by the New York Post objecting to Facebook's labeling of an opinion piece by New York Post contributor Steven W. Mosher as containing misleading information. As explained in a previous review by Health Feedback, Mosher linked COVID-19 with bioweapons research based on false and misleading statements that are not supported by current evidence. On 30 April, the U.S. Office of the Director of National Intelligence announced that "The Intelligence Community also concurs with the wide scientific consensus that the COVID-19 virus was not man made or genetically modified".

The article also includes an entire post from 12 June by Old-Thinkers News which states that "Gates and the World Bank helped build a global digital ID structure before the COVID-19 pandemic". The post is referring to ID2020, or the Digital Identity Alliance which was launched in 2016 by UNICEF, the World Bank, and the Bill and Melinda Gates Foundation. ID2020 seeks to provide digital identification to all citizens who desire it in order for them to obtain equal access to services, while improving privacy protections and giving them control over their information. However, the post uses the existence of ID2020 to make the unsupported claim that Gates intends to conduct global surveillance using COVID-19 vaccines.

This conspiracy theory mixes up two unrelated initiatives funded by the Gates Foundation: digital identity and vaccines. According to the ID2020 alliance, their aim is to grant access to essential goods and services in regions where national identification systems are not possible, such as developing countries. None of the technology involved in the initiative allows for the tracking or surveillance of citizens. The idea for this conspiracy theory comes from a Reddit thread in which Gates mentioned the possibility of a digital certificate for COVID-19 immunity.

The article also reproduces an 8 April Instagram post from Robert F. Kennedy, Jr. containing false claims about vaccination programs that were funded by the Gates Foundation and conducted in partnership with the World Health Organization (WHO) and the Program for Appropriate Technology in Health (PATH). Specifically, Kennedy states that different vaccines against polio, human papillomavirus (HPV), malaria, meningitis, and tetanus, have caused severe side effects and deaths in Africa and India. Each one of these claims is reviewed below. The nephew of the former U.S. president John F. Kennedy and leader of the anti-vaccine group the World Mercury Project, Robert F. Kennedy, Jr. is a popular source of vaccine-related misinformation, some of which have been reviewed by Health Feedback here, here, and here.

The claim that polio vaccines in India "paralyzed 496,000 children between 2000 and 2017", has been widely debunked by AFP, Lead Stories, and PolitiFact. The oral poliovirus vaccine (OPV) contains an attenuated strain of the virus that can be excreted from the body while immunity is building. On rare occasions, the virus can mutate into vaccine-derived polioviruses (VDPVs) which are capable of causing paralysis. In populations with inadequate sanitization and low immunization coverage, an excreted virus may persist for long periods of time and start circulating among individuals (cVDPVs).

The oral polio vaccine does not cause polio in those who receive the vaccine, and immunization protects against both vaccine-derived and wild polioviruses. Hence cVDPV outbreaks occur only in underimmunized populations among individuals who are not vaccinated, as explained in another Health Feedback review. As Dr Pascal Mkanda, head of the WHO's Polio Eradication Programme, explained in this November 2019 article, the solution to cVDPV is not to stop vaccinating, but to improve vaccine coverage. Since 2000, India has reported a total of 5,468 polio cases, only 17 of which were due to cVDPVs. After the last cases in 2010, the country was officially declared polio-free in 2014.

Kennedy's statement that "by 2018, ¾ of global polio cases were from Gates' vaccines" is also highly misleading. In 1988, when the Global Polio Eradication Initiative was launched, an estimated 350,000 children were paralyzed by polio each year. Although about 70% of polio cases worldwide in 2019 were due to cVDPV outbreaks, it is good to keep in mind that the total number of polio cases worldwide in 2019 stood at 539, a stark contrast to 350,000 in 1988. This massive reduction in polio cases has been due to polio vaccination. Overall, vaccination campaigns have eradicated polio from the vast majority of developing countries, preventing an estimated 16 million cases and 1.5 million deaths worldwide.

Kennedy also claims in his post that a 2009 "experimental" vaccine against HPV in India caused seven deaths, along with autoimmune and fertility disorders in 1,200 girls. The vaccines he refers to, Gardasil and Cervarix, were far from experimental. They had already been approved in the U.S. and other countries years earlier. In fact, the aim of the 2009 study, conducted by PATH, was not to test the safety or efficacy of the vaccines, but to validate the use of cost-effective vaccines to reduce human papillomavirus (HPV) infections in low-income communities. As previously reviewed, extensive international studies confirm an excellent safety profile of the HPV vaccine with no serious side effects. In contrast, HPV-related cervical cancer killed 311,000 women in 2018, 85% of them in developing countries, and particularly in India.

The deaths of the seven girls in the PATH study were investigated and found to be unrelated to the vaccines, as previously covered by PolitiFact, Reuters, and Snopes. The Indian government absolved the organizations involved in the HPV vaccination study of all responsibility, and never sued or banned PATH or the Gates Foundation from operating in India, as it has been suggested. However, PATH did receive a warning for inadequately handling ethical consent.

The post also inaccurately claims that three different vaccination programs funded by the Gates Foundation in Africa caused severe side effects or deaths. The first program was a clinical trial for a malaria vaccine that Kenedy claims killed 151 African infants and caused "paralysis, seizure, and febrile convulsions to 1,048 of the 5,949 children". This statement is a complete misrepresentation of the results published in The New England Journal of Medicine in 2011[1]. To control for side effects of the malaria vaccine, adverse events and all-cause deaths were compared to two non-malaria vaccination groups. Only ten children manifested vaccine-related effects, none of them paralysis, and all recovered. Deaths were equally distributed in all groups and were due to causes unrelated to the vaccine. The study concluded that the vaccine reduced malaria in children by half of the incidence in the previous years. The effect of the vaccine persisted for one year without severe adverse effects.

The second program that Kennedy mentions is the 2002 MenAfriVac meningitis vaccine campaign, during which he claims that the vaccine caused paralysis in 50 children in Sub-Saharan Africa. This was reviewed by Africa Check and found to be false. After a complete vaccination campaign in Chad, a single town reported unusual reactions to the vaccine in 35 children and youth. Later, an unvaccinated child complained of the same symptoms. An investigation by a group of independent experts concluded that the events were not related to the vaccine, but to a mass psychogenic phenomenon. A 2013 study published in The Lancet showed that the vaccination campaign was highly effective and reduced the cases of meningitis in Chad by 94%[2].

Finally, Kennedy repeats the old myth that the tetanus vaccine contains a sterilizing formula, a claim that the WHO has already debunked. Africa Check and Snopes have explained that the origin of this claim is a small clinical trial conducted in India in 1994[3]. This study aimed to develop a birth control vaccine using human chorionic gonadotropin (hCG), a hormone necessary for pregnancy, coupled to a modified tetanus toxin as a carrier. This trial was wrongly linked to the first tetanus vaccination campaigns in Kenya, and the rumor of involuntary sterilization using tetanus vaccines began to spread. The use of pregnancy test kits, which were not suitable for assessing the presence of hCG in the vaccine, led to false positives that fueled the rumors. However, laboratory analysis using a more suitable technique found no hCG in the tetanus vaccine.

The claim that vaccines are being used to sterilize the population was linked to Gates after he stated in a CNN interview in 2011 that vaccines "could reduce population growth". As explained by Snopes, Gates was referring to the idea that population growth can be slowed not by birth control, but by reducing childhood mortality. The Foundation's Annual Letter explains that there is a correlation between infant mortality rates and fertility rates and, as the number of kids who survive to adulthood increases, parents tend to have fewer children. This theory is still a subject of debate among the scientific community.

To support his claims, Kennedy cherry-picks an observational study published in EBiomedicine in 2017 to show that the diphtheria-tetanus-pertussis (DTP) vaccine increases child mortality[4]. The data analyzed is from immunizations in the early 1980s using the whole-cell version of the pertussis vaccine. The acellular version of the vaccine (DTaP) which is currently available provides shorter protection, but has fewer side effects. In addition, the study introduced an important bias by using a rather small sample size and including all-cause mortality events, many of them unrelated to the vaccine. More recent and extensive studies show that DTP vaccines are safe and effective in preventing disease." (Carballo-Carbajal 2021)

To demonstrate the behaviors of Robert F. KennedyJr., will visit factcheck.org to help us better understand how he rehashes arguments and misrepresents data. This is critical in helping us understand the root of the problem when it comes to the disinformation that has cultivated a mindset that has driven people to believe and not know.

> RFK Jr. Video Pushes Known Vaccine Misrepresentations

Several vaccine falsehoods and misrepresentations have been strung together in a video aimed at discouraging Black people from getting vaccinated against COVID-19.

The hourlong video, called "Medical Racism: The New Apartheid," is hosted by Robert F. Kennedy Jr.'s anti-vaccination organization, Children's Health Defense. It includes mostly rehashed claims about vaccine safety framed to exploit distrust of the medical establishment in Black communities.

The video, which was made available on March 11, doesn't offer any evidence that COVID-19 vaccines are harmful. Instead, it relies on innuendo — citing historical examples of ethical failures in medicine, misrepresenting various scientific studies, and suggesting that the medical establishment can't be trusted.

For example, Kennedy — who is a lawyer, not a doctor — ends the video by discouraging viewers from following the advice of public health officials, like Dr. Anthony Fauci, director of the National Institute of Allergy and Infectious Diseases. "Don't listen to me, don't listen to Tony Fauci, and don't listen to your doctor," Kennedy says before advising viewers to review the package inserts for vaccines and question whether the ingredients are safe. (See SciCheck's articles on each vaccine: "A Guide to Johnson & Johnson's COVID-19

Vaccine," "A Guide to Moderna's COVID-19 Vaccine" and "A Guide to Pfizer/BioNTech's COVID-19 Vaccine.")

Likewise, promotional material for the video baselessly claims that "[t]here are signs that history is repeating itself with the coronavirus" and questions, "Is the COVID-19 Vaccine Safe?"

This is similar to the approach taken in another disinformation video we recently wrote about. In both cases, the videos acknowledge that Black people in the U.S. are suffering higher rates of illness and death from COVID-19, while simultaneously casting doubt on the vaccines that could curb the virus.

Referring to Kennedy, Dr. Richard Allen Williams, founder and president of the nonprofit Minority Health Institute, told us in a phone interview: "It seems what he does is take old information and make a … leap to the present time and say what happened then is what is applicable to what is going on now."

But historical examples of unethical medical conduct are very different from the authorized COVID-19 vaccines.

All three available vaccines in the U.S. went through clinical trials with tens of thousands of participants before the Food and Drug Administration granted them each an emergency use authorization. About 10% of participants in those trials were Black for two of the vaccines and about 17% were Black for the third. The trials were overseen by independent data and safety monitoring boards, and the results were reviewed by the FDA and an outside panel of experts. These tested, authorized vaccines are now available to the general population.

Still, disinformation about the vaccines persists.

For example, Williams — also a clinical professor of medicine at the UCLA School of Medicine — noted that when baseball legend and civil rights activist Henry "Hank" Aaron died of natural causes on Jan. 22, more than two weeks after he received a COVID-19 vaccine,

Kennedy "seemed to make an association between Hank's death and his previous vaccination." We wrote about the misleading comments made by Kennedy and his group at the time.

"That's an example of the dangerous thing he does," Williams said, referring to the "harmful effects of influencing people to not take the vaccine."

Rehashed Claim About CDC Study, Autism

The video misrepresents a 2004 study by the Centers for Disease Control and Prevention that did not find an association between autism and measles, mumps and rubella vaccines.

The study, which was published in the journal Pediatrics, looked at the age of children living in the Atlanta area when they received their first MMR vaccine and whether there was a correlation between the timing of the vaccine and a diagnosis of autism. If vaccines were contributing to autism, then one might expect to see more autism cases in kids who received the vaccine earlier, as we wrote about a similar claim.

The study did not find that vaccines were causing autism.

But Kennedy's organization falsely claims that the "study discovered that African-American boys who receive the MMR vaccine 'on-time' by the age of 3 are 3.36 times more likely to be diagnosed with severe autism as Black boys who waited until they were older."

That's not what the study found. That's what a purported reanalysis of the data claimed to find a decade later.

Brian Hooker — who has a degree in chemical engineering, not medicine, and has contributed articles to the Children's Health Defense website — wrote the reanalysis for the journal Translational Neurodegeneration. But the paper was retracted just over a month after it was published.

It's worth noting, too, that his research was funded by Focus Autism, an anti-vaccine group now known as Focus for Health, where Hooker reportedly served on the board.

Also, Hooker had filed a claim under the National Vaccine Injury Compensation Program alleging that his son's autism was caused by vaccines. The lawsuit was pending at the time Hooker published the paper, but was later dismissed in a 58-page opinion in which the judge said, "[T]his case is not a close call." The judge found no connection between vaccines and the boy's disorder.

The paper was retracted because "there were undeclared competing interests on the part of the author" and there were also "concerns about the validity of the methods and statistical analysis."

Prior to publishing his paper, Hooker had spoken to Dr. William Thompson, one of the researchers on the original CDC study. Thompson had shared his concerns about the study's process, but later explained that he did not know Hooker had been recording their conversations and had no control over how Hooker used the recordings.

Thompson still works at the CDC, and when we reached him by phone, he referred us to his lawyer, Rick Morgan. Neither would speak to us on the record, but Morgan provided us with Thompson's 2014 statement. It said, in part: "I want to be absolutely clear that I believe vaccines have saved and continue to save countless lives. I would never suggest that any parent avoid vaccinating children of any race. Vaccines prevent serious diseases, and the risks associated with their administration are vastly outweighed by their individual and societal benefits."

So, the claim in the video uses a discredited paper to misrepresent the findings of a CDC study.

Misrepresented Mayo Clinic Study

In another distortion of scientific research, the video twists the meaning of a preliminary 2014 study that found some Somali Americans developed twice the antibody response to rubella after getting the vaccine compared with Caucasians.

The Children's Health Defense material, however, claims that the study, which was done by Dr. Gregory Poland of the Mayo Clinic, "means that African-American children are being 'overdosed' with today's current vaccine concentration."

But that's not at all what the study found, Poland told us in a phone interview.

He explained that this claim is based on a hypothesis generating study, which means that researchers look at a small sample, make observations and speculate about what those observations could mean. Then they do a larger study.

In this case, Poland said, "We haven't done the larger study."

In 2014, Poland and a team of researchers at the Mayo Clinic looked at 1,100 healthy children and young adults in Rochester, Minnesota, and more than 1,000 participants from the U.S. Naval Health Research Center in San Diego as the control group, as well as a group of recent immigrants to Rochester from Somalia.

As we said, the Somali Americans developed twice the antibodies to rubella after getting the vaccine as their white counterparts.

Since they haven't done the larger study, researchers don't know why there is a difference or what it might mean. It could mean that white children are being underdosed, Poland said, explaining that without further study it's unclear.

What is clear is that the claim from the Children's Health Defense is wrong. "We do not have a study that shows African Americans need half the dose. We do not have a study that shows African American children are being overdosed," Poland said.

He described the claim as being "like a good conspiracy theory — it contains a grain of truth with a lot of speculations around it."

The grain of truth is that researchers found higher levels of antibodies in Somali immigrants, but the rest is conjecture. The higher levels may not have anything to do with the vaccine, Poland said. For example, the study subjects could have been exposed to rubella in refugee camps before arriving in the U.S., so the antibodies from the vaccine would have added to what was already in their systems.

So, again, this claim is a misrepresentation of a preliminary study.

Understanding Flawed Measles Vaccines Trial

The video also revisits a controversial clinical trial from 30 years ago. Several years after the study concluded, the CDC acknowledged it had failed to notify parents that one of the vaccines used in the study was experimental, and of the specific risks involved.

"The Tuskegee experiment that took place from 1932 to 1972 — the CDC promised us that an experiment like this would never happen again, but yet in 1989 there was a Los Angeles Times article that came out that revealed the startling truth: that the CDC admitted that they experimented on 1,500 Black and Latino boys in Los Angeles — without the parents knowing that this was an experimental test," Tony Muhammad, of the Nation of Islam, says in the video.

That's referring to a clinical trial for measles vaccines that the CDC conducted in Los Angeles on babies up to a year old between 1990 and 1991 — amid a measles epidemic that began in 1989.

The CDC acknowledged in a 1996 Los Angeles Times story that the families involved in the Los Angeles trial were not informed that one of the vaccines used in the trial was experimental. The majority of the children were African American and Latino, the newspaper reported.

The study involved about 1,500 children, who received either high or standard doses of the Edmonston-Zagreb, or EZ, vaccine or standard doses of the Moraten vaccine, according to testimony to Congress in

May 1997 by then-CDC Director Dr. David Satcher. The Moraten vaccine was already licensed for use in the U.S. during the trial. But while other countries widely used standard doses of the EZ vaccine at the time, it was new to the U.S., and the high-dose version was experimental.

The CDC stopped the trial in October 1991 after separate studies in Senegal and Haiti suggested baby girls vaccinated with the higher potency version were at increased risk of dying in the two to three years following vaccination.

Satcher, who was not director when the trial took place, said in his prepared remarks that the EZ vaccines that were used in the trial were approved for investigation by the Food and Drug Administration but that the consent form for families was "deficient."

He said a subsequent review by the Office for Protection from Research Risks (now the Office for Human Research Protections in the Department of Health and Human Services) determined the study was "scientifically and ethically justified," but the consent form failed to identify "the EZ vaccine as experimental" and failed to provide an "adequate description of the foreseeable risks of the experimental EZ vaccine."

According to the Los Angeles Times, "the form did not say that E-Z was experimental or unlicensed. But a brochure that accompanied the form said: 'This vaccine has been shown to be effective in younger children. Over 200 million children around the world have received this vaccine, but Los Angeles County is the first place in the United States where it is being offered.'"

Satcher said in a follow-up effort with families, "parents were informed of the reason for the follow-up study, including the fact that some studies had found lower survival in those children who received high potency vaccine."

The CDC didn't find any ill effects associated with the high potency EZ vaccine in its study. It did not identify any children who took part in the Los Angeles study and received the high potency EZ vaccine who "suffered a significant health problem that can be associated with the vaccine," Satcher said.

One child died about a year after receiving a standard-dose EZ vaccine, according to Satcher, and "experts reviewed the death certificate, the circumstances surrounding the death, and the autopsy report and all agreed with the conclusion that the death was in all likelihood unrelated to the vaccine."

Williams, of the Minority Health Institute, told us that the mistakes made were legitimate issues that shouldn't be forgotten. But the effort to compare that situation to the COVID-19 vaccine was "a matter of apples and oranges," he said.

"It's not logical to take the results of what happened in an entirely different context and apply it to our current situation," added Williams, a past president of the National Medical Association, which is the largest and oldest association of Black physicians.

It's worth emphasizing, as we said earlier, that the COVID-19 vaccines authorized in the U.S. have
successfully undergone clinical trials that demonstrated their safety.

Williams said that he believes the issues with the measles vaccines trial weren't given sufficient attention at the time. But, he said, that doesn't substantiate suggestions that the same thing is happening now.

"It needs to be understood, there was a mistake made," Williams said, in order to make sure "those kinds of mistakes are not being made" again. "In fact, there are all sorts of preventive measures" to ensure that kind of mistake doesn't occur, he added.

Satcher outlined some changes that were made following the trial during his 1997 testimony, including institutional review changes and the creation of a checklist of elements that are to be included on consent forms.

Revisiting the Impact of Vitamin D

The video also revisits the topic of vitamin D and COVID-19. It presents a clip of Fauci speaking in an Instagram interview, telling the actress Jennifer Garner: "If you are deficient in vitamin D that does have an impact on your susceptibility to infection — so I would not mind recommending, and I do it myself, taking vitamin D supplements."

Fauci was speaking (at about 31:30) about infections in general, however — not the novel coronavirus specifically, as the video may imply.

"My question is, why isn't this being discussed, or why isn't the CDC or the World Health Organization jumping on this as a potential solution to do more research around it?" Dr. Charles Penick, a family medicine physician, asks in the video.

It's worth noting that there are many studies under way looking into the relationship between vitamin D and COVID-19.

As we've explained before, while vitamin D might be helpful in terms of COVID-19, there isn't enough scientific evidence yet to know whether it can treat or prevent the disease. That said, health experts do advise getting enough vitamin D as part of a healthy lifestyle, regardless. Most of a person's vitamin D is made in the skin by exposure to sunlight.

One study published in October in the Journal of Clinical Endocrinology & Metabolism, which evaluated the vitamin D levels of 216 hospitalized COVID-19 patients, found that the hospitalized patients had lower levels of vitamin D compared with a group in the general population — although it didn't identify a relationship between lower vitamin D levels and more severe disease among those patients.

While suggestive, the study doesn't prove that vitamin D was a factor in whether those patients fell ill or that taking vitamin D supplements would have helped, as it was not a randomized controlled trial.

Two other recent studies — which have not yet been peer-reviewed — appear to cast doubt on the effectiveness of vitamin D in preventing COVID-19. (Spencer & Finchera 2021)

Now you see why Robert F. Kennedy, Jr., is so dangerous the amount of claims he's made over the years only to re-hash those claims over and over and add something new to them as if it will finally bring clarity to his argument and it still does not, but that was just a sample size of his rhetoric because in a more recent claim regarding the death of Hank Aaron he misleads people again: FULL

CLAIM: "Aaron's tragic death is part of a wave of suspicious deaths among elderly closely following administration of COVID vaccines." Misleading: The statement implies that the COVID-19 vaccines were the cause of a "wave of suspicious deaths" among elderly people. There is no evidence available to support this claim. Simply because an event occurred after vaccination does not mean that vaccination caused the event. In order to establish whether a vaccine caused an adverse event, one has to look beyond anecdotes and compare the incidence rate of the adverse event between a group that received the vaccine and a group that didn't receive the vaccine. (Teoh 2021)

Joseph Mercola is another alternative medicine pusher who is an osteopathic physician. He is responsible for this whole vitamin C, and D have been adopted as treatments for Covid. "Despite a lack of evidence that vitamins are effective against the novel coronavirus, a doctor with a history of making misleading claims said they are used as a treatment for the virus. An April 7 article on the website of the Organic Consumers Association by Joseph Mercola headlined "Vitamins C and D finally adopted as coronavirus treatment" claims that "vitamins C and D are now (finally) being adopted in the conventional treatment of novel coronavirus." Mercola is a doctor of osteopathy who promotes alternative medicines. The U.S. Food and

Drug Administration has issued Mercola at least three warning letters accusing him of making "false or misleading claims" about products he promoted on his website. For years, medical experts have criticized Mercola for sharing dangerous information. "The information he's putting out to the public is extremely misleading and potentially very dangerous," Dr. Stephen Barrett told Chicago Magazine in 2012 in an article about Mercola. Barrett runs QuackWatch.org, a medical watchdog website. "He exaggerates the risks and potential dangers of legitimate science-based medical care, and he promotes a lot of unsubstantiated ideas and sells (certain) products with claims that are misleading." Mercola's claim about vitamins and the coronavirus cites a New York Post article from March 24 that describes the use of vitamin C by Northwell Health, a New York hospital system, to treat patients with coronavirus. Northwell spokesperson Jason Molinet confirmed to USA TODAY that "vitamin C was one of many therapies employed at the discretion of physicians in our health system." Molinet declined to answer follow-up questions about how widespread the use of vitamin C was, what the results of the treatment were and what studies or data Northwell relied on when deciding whether to use vitamin C as part of COVID-19 treatment. He declined to make a doctor available to speak about the treatment, saying, "That's the extent of our statement on this." "There is no evidence that taking extra vitamin C will fight against COVID-19. In fact, the body can only absorb a certain amount of vitamin C, and any excess will be excreted," the National Foundation for Infectious Diseases says in a graphic on its website. The Centers for Disease Control and Prevention and the World Health Organization said the only way to minimize the chances of contracting the virus is to take preventive steps such as social distancing from other people, frequent hand-washing and cleaning of often-used surfaces. USA Today fact checker concludes: Though vitamin C is used, at least in one New York hospital system, to help treat some patients on a case-by-case basis, there is no known evidence to suggest it is effective. Occasional use of vitamins C or D in COVID-19 treatment at the discretion of a patient and doctor is not the same as saying they are being adopted "in the conventional treatment" of the coronavirus, as Mercola's article says. (Gruber-Miller 2020)

Published in Jama Network a study on Zinc, Vitamin D, and Vitamin C provided testing results: Results A total of 214 patients were randomized, with a mean (SD) age of 45.2 (14.6) years and 132 (61.7%) women. The study was stopped for a low conditional power for benefit with no significant difference among the 4 groups for the primary end point. Patients who received usual care without supplementation achieved a 50% reduction in symptoms at a mean (SD) of 6.7 (4.4) days compared with 5.5 (3.7) days for the ascorbic acid group, 5.9 (4.9) days for the zinc gluconate group, and 5.5 (3.4) days for the group receiving both (overall P=.45). There was no significant difference in secondary outcomes among the treatment groups. Conclusions and Relevance In this randomized clinical trial of ambulatory patients diagnosed with SARS-CoV-2 infection, treatment with high-dose zinc gluconate, ascorbic acid, or a combination of the 2 supplements did not significantly decrease the duration of symptoms compared with standard of care. (Thomas MD MS, Patel MD MS, Bittel BSN RN 2021)

The second study present by Jama Network stated: Results Of 240 randomized patients, 237 were included in the primary analysis (mean [SD] age, 56.2 [14.4] years; 104 [43.9%] women; mean [SD] baseline 25-hydroxyvitamin D level, 20.9 [9.2] ng/mL). Median (interquartile range) length of stay was not significantly different between the vitamin D3 (7.0 [4.0-10.0] days) and placebo groups (7.0 [5.0-13.0] days) (log-rank P=.59; unadjusted hazard ratio for hospital discharge, 1.07 [95% CI, 0.82-1.39]; P=.62). The difference between the vitamin D3 group and the placebo group was not significant for in-hospital mortality (7.6% vs 5.1%; difference, 2.5% [95% CI, −4.1% to 9.2%]; P=.43), admission to the intensive care unit (16.0% vs 21.2%; difference, −5.2% [95% CI, −15.1% to 4.7%]; P=.30), or need for mechanical ventilation (7.6% vs 14.4%; difference, −6.8% [95% CI, −15.1% to 1.2%]; P=.09). Mean serum levels of 25-hydroxyvitamin D significantly increased after a single dose of vitamin D3 vs placebo (44.4 ng/mL vs 19.8 ng/mL; difference, 24.1 ng/mL [95% CI, 19.5-28.7]; P<.001). There were no adverse events, but an episode of vomiting was associated with the intervention. Conclusions and Relevance: Among hospitalized patients with COVID-19, a single high dose of vitamin D3, compared with placebo, did not significantly reduce hospital length of stay. The findings do not support the use of a high dose of vitamin D3 for treatment of moderate to severe COVID-19. (Murai Phd, Fernandes Phd, Sales MSc 2021)

Another claim made by Mercola's: "CLAIM: People may be more susceptible to serious COVID-19 illness after they have been vaccinated. AP'S ASSESSMENT: False. Research has shown that the Pfizer and Moderna vaccines have been proven to be 95% effective in preventing COVID-19 illness. Experts say there is "abundant" evidence that people who get shots will not become more sick should they later get the virus. THE FACTS: A online post contains false information by suggesting that people who receive the COVID-19 vaccine may experience more severe symptoms if they are exposed to the virus." (Associated Press 2021)

The Dirty Dozen have a few things in common most of them are osteopathic physicians the others are public speakers and influencers, but they all promote and push alternative medicines without evidence that supports their claims. Yet millions of people subscribe to their ideology. In the middle of a pandemic with millions of lives lost simply pushing a false alternative is not a solution to the bigger problem. Each person only looks to poke holes in the only data scientist have provided the public. As a researcher I decided to provide you with a list of claims not used in the CCDH. Although it is essential I wanted to demonstrate how Social Media post from the Dirty Dozen expand beyond the claims listed in CCDH article. We will now turn our attention to numerous social media claims inspired by the loudest anti-vaxxers.

Then we have Trump, during the end of his presidency states "President Donald Trump said Monday that he has been taking anti-malaria drug hydroxychloroquine for over a week to prevent coronavirus." (Lovelace & Bruninger 2020) He continued to tout this anti-malaria drug as a preventative. He stated this without providing any scientific data that would prove his position, however studies would follow proven that hydroxychloroquine was useless in prevention and treatment. Many herbalist also argued this point all over social media sites. In an study published by Jama Network late November 2020 states: "Results Among 479 patients who were randomized (median age, 57 years; 44.3% female; 37.2% Hispanic/Latinx; 23.4% Black; 20.1% in the intensive care unit; 46.8% receiving supplemental oxygen without positive pressure; 11.5% receiving noninvasive ventilation or nasal high-flow oxygen; and 6.7% receiving invasive mechanical ventilation or extracorporeal membrane oxygenation), 433 (90.4%) completed the primary outcome assessment at 14 days and the remainder had clinical status imputed. The median duration of symptoms prior to randomization was 5 days (interquartile range [IQR], 3 to 7 days). Clinical status on the ordinal outcome scale at 14 days did not significantly differ between the hydroxychloroquine and placebo groups (median [IQR] score, 6 [4-7] vs 6 [4-7]; aOR, 1.02 [95% CI, 0.73 to 1.42]). None of the 12 secondary outcomes were significantly different between groups. At 28 days after randomization, 25 of 241 patients (10.4%) in the hydroxychloroquine group and 25 of 236 (10.6%) in the placebo group had died (absolute difference, −0.2% [95% CI, −5.7% to 5.3%]; aOR, 1.07 [95% CI, 0.54 to 2.09]). Conclusions and Relevance Among adults hospitalized with respiratory illness from COVID-19, treatment with hydroxychloroquine, compared with placebo, did not significantly improve clinical status at day 14. These findings do not support the use of hydroxychloroquine for treatment of COVID-19 among hospitalized adults." (Self MD MPH, Semler MD, Leither DO 2020)

The NIH haunted its trial concluded that the "study shows treatment does no harm, but provides no benefit." (NIH 2020) Because the sitting president at the time made a claim science sat out to confirm it and realized more than once that this anti-malaria drug served no purpose in helping treat or prevent Covid-19.

The Claim: Can mRNA vaccines alter your DNA? "Social media users have been sharing articles that claim Moderna's chief medical officer Tal Zaks has said mRNA vaccines – like the Moderna vaccine for COVID-19 – alter DNA. The Facts: This claim is false. It is based on comments made by Zaks that have been misconstrued, and Reuters has found no evidence of him making any such comments elsewhere." (Reuters 2020)

At the beginning of the pandemic people blamed technology for the spread of a novel virus, and the poster child was non other than Bill Gates for his contributions. The Claim: Did Bill Gates plan to use 5G technology to plan a pandemic? The Facts: "Facebook users have been sharing a video that makes multiple false claims about COVID-19, for instance that Bill Gates planned the pandemic and 5G technology was involved in its spread." (Reuters 2020) This claim has been debunked numerous times yet it has re-appeared every full moon. You can't use technology to spread a virus that is not how viruses work.

This claim has lead to conspiracist regurgitating it over and over and using a photo to assume that this gentleman was even a Doctor. The Claim: Covid-19 is a hoax? A photo showing a doctor standing in front of empty hospital beds at a Reno, Nevada, auxiliary care site for COVID-19 patients proves that the coronavirus pandemic is a hoax.

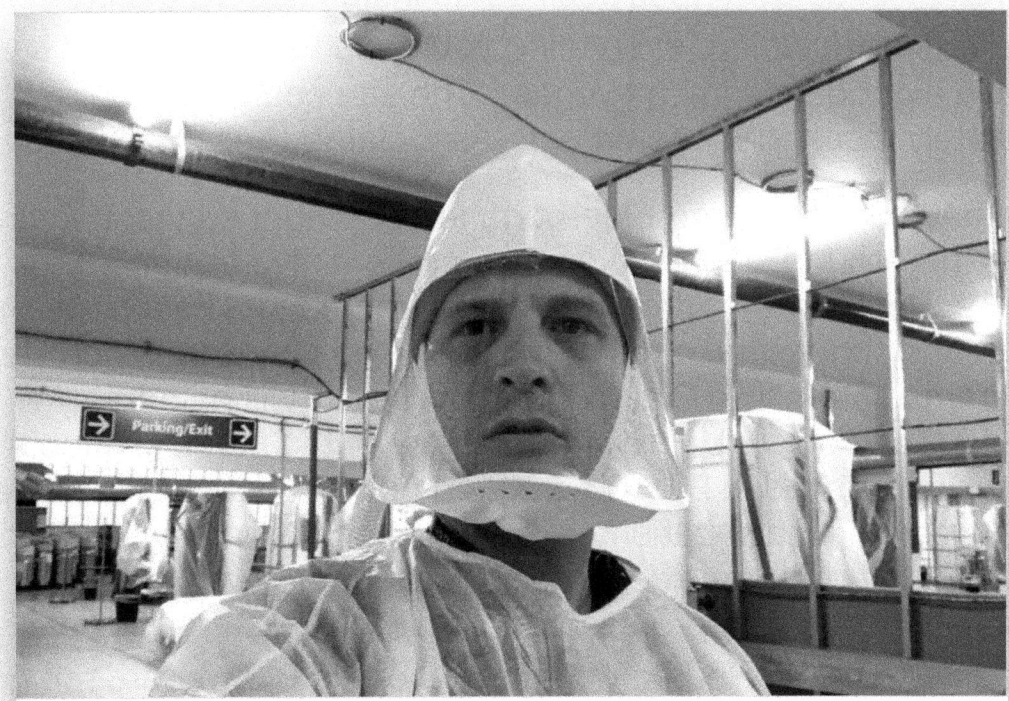

This Nov. 12, 2020 selfie photo provided by the Renown Regional Medical Center shows Dr. Jacob Keeperman, the Renown Transfer and Operations Center medical director who snapped the photo on the opening day of the Renown Regional Medical Center's alternative care site located in a parking garage. Keeperman's photo was being misrepresented online to support the false narrative that the coronavirus pandemic is a scam. (Jacob Keeperman/Renown Regional Medical Center via AP)
THE ASSOCIATED PRESS

AP'S ASSESSMENT: False. The photo was taken the day the alternative care site was opened, and patients had yet to arrive. Renown Regional Medical Center said the site, which is housed in a parking garage, has treated 198 coronavirus patients since it first opened. THE FACTS: In recent weeks, social media posts have shared a variety of falsehoods about the hospital's parking garage site, with some posts saying that visitors went there and found no patients, which they then cited as evidence that the virus is a hoax. President Donald Trump propelled the misinformation Tuesday, retweeting the photo of the doctor amid empty beds to his more than 80 million followers. "Fake election results in Nevada, also!," he said of the tweet suggesting that the parking garage site and pandemic were both fake. According to Renown hospital officials, the alternate care site in the parking structure currently has 42 patients and has served 198 patients since opening day in November. The site, which was set up for patients who do not require long-term care, can house more than 1,400 patients. The Twitter account making the accusations about the Renown facility has repeatedly criticized the state's governor for his coronavirus restrictions. The account @Networkinvegas describes itself as an inside source of information on Las Vegas, including "everything you need to network, hook up, and have a good time in Las Vegas." The site could not be immediately reached for comment. (Associated Press 2020)

People have been burying Elderberry left and right suggesting and posting this around social media as a preventative treatment for Covid-19. Recently this information was shared in a Facebook thread by someone suggesting a recently diagnosed Covid patience should take it immediately. The Claim: Elderberry daily will help treat Covid-19. The Facts: "Unproven. A review of the medical effects of elderberry published in 2014 (here) did not find conclusive evidence that it helped influenza patients. There has not been any significant research into the use of elderberry to prevent or cure COVID-19. Sheena Cruickshank, a professor of immunology at the University of Manchester, UK told Reuters: "Vitamin C or elderflower are not going to specifically help. Elderflower has vitamin C, but vitamin C is a water soluble vitamin so we don't store it. It's an essential vitamin for health but not necessarily a virus buster." (Reuters 2020)

The Claim: "Covid-19 is NOT killing people. Weak immune systems and bad doctors are." The Facts: "While the Centers for Disease Control and Prevention found that 94% of people who died of COVID-19 had other conditions, they still died from COVID-19." (Rourke 2020)

This claim is a very popular claim recently I saw it shared on someones post on social media who recently shared that they had tested positive for Covid-19. The Claim: The mRNA vaccines change your DNA and could cause cancer. The Facts: None of the vaccines interact with or alter your DNA in any way, and therefore cannot cause cancer. Messenger RNA (mRNA) is not the same as DNA and cannot be combined with DNA to change your genetic code. Here's now mRNA vaccines actually work: The mRNA vaccines use a tiny piece of the coronavirus' genetic code to teach your immune system how to make a protein that will trigger an immune response if you get infected. The mRNA is fragile, so after it delivers the instructions to your cells, it breaks down and disappears from the body (in about 72 hours). The mRNA never even goes into the nucleus of the cell — the part that contains your DNA. Therefore, there is no truth to the myth that somehow the mRNA vaccine could inactivate the genes that suppress tumors. (MSKCC 2021)

Candace Owens tweeted on April 6, 2020 The Claim: Apparently, doctors and nurses around the world are wondering why no one is dying from heart attacks and strokes anymore. Flu and pneumonia deaths also went off a cliff. Turns out everyone is only dying of #Coronavirus now. Gee. I wonder why. The Facts: "Patients who test positive for the coronavirus are likely being included in nationwide death counts. But doctors say that's actually an undercount because of a lack of available testing, among other factors. Coronavirus is more difficult for people with pre-existing heart and lung problems, which could lead to respiratory or cardiac arrest." (Greenberg 2020)

Facebook post claims: Covid-19 is man made. The Facts: AP'S ASSESSMENT: False. Scientists say the molecular structure of SARS-CoV-2 rules out the possibility that the virus was created in a lab. (Associated Press 2020)

Phil Valentine in a lecture on YouTube advised people at the earlier onset of the Pandemic to paint your houses with lead paint. The Claim: Lead paint will stop the Coronavirus from entering into your home. The Facts: Lead paint is still present in millions of homes, sometimes under layers of newer paint. If the paint is in good shape, the lead paint is usually not a problem. Deteriorating lead-based paint (peeling, chipping, chalking, cracking, damaged, or damp) is a hazard and needs immediate attention. Lead-based paint may also be a hazard when found on surfaces that children can chew or that get a lot of wear-and-tear, such as: Windows and window sills; Doors and door frames; and Stairs, railings, banisters, and porches. (EPA 2020) There is no evidence that suggest that a hazardous substance like Lead and prevent you from being infected with a virus.

Another regurgitated fear that was mentioned often before Covid restrictions began to ease was-The Claim: that mask did not help prevent transmission of Covid-19 nor did social distancing. The Facts: "Scientific evidence is clear: Social distancing and wearing a mask help prevent people from spreading COVID-19, and masks also protect wearers from being infected themselves, two UC Davis Health experts said on UC Davis LIVE: COVID-19. Along with preventing someone from transmitting the coronavirus, a range of new research shows that the risk of infection to the wearer is decreased by 65%, said Dean Blumberg, chief of pediatric infectious diseases at UC Davis Children's Hospital. (UC Davis Health 2020)

The Claim: Social media users have claimed the presence of lipid nanoparticles in a COVID-19 vaccine means it could contain small robots or computers. The Facts: "This is false - these nanoparticles are tiny lipid droplets that transport and protect the vaccine component. The term "nano", however, is simply a unit of size. Nanotechnology can refer to science conducted at the nanoscale of about 1 to 100 nanometers. Similarly, the general definition of "nanoparticle" is a small particle that is between 1 and 100 nanometres in size. And this in case, the term "nanoparticle" refers to a tiny lipid droplet that carries the vaccine component. Lipids are substances that are not soluble in water, like fats." (Reuters 2020)

The claim: Tennessee Nurse passes out after getting Covid-19 vaccine. In a video shared thousands of times we see a nurse receiving the vaccine and then suddenly faint. The video shared is not played in its full length and disregards the nurse answering questions after her brief experience. The Facts: True she did pass out she also stated 2 minutes after her passing out that she suffers from Vasovagal syncope. "Vasovagal syncope (vay-zoh-VAY-gul SING-kuh-pee) occurs when you faint because your body overreacts to certain triggers, such as the sight of blood or extreme emotional distress. It may also be called neurocardiogenic syncope. The vasovagal syncope trigger causes your heart rate and blood pressure to drop suddenly. That leads to reduced blood flow to your brain, causing you to briefly lose consciousness. Vasovagal syncope is usually harmless and requires no treatment. But it's possible that you may injure yourself during a vasovagal syncope episode. Your doctor may recommend tests to rule out more-serious causes of fainting, such as heart disorders." (Mayo Clinic 2021)

Post shared all over Social Media about vaccine injection photos created a lot of noise. People interjected that lawmakers such as Nancy Pelosi and front line workers where not receiving the vaccine. The claim: Image with cap over syringe shows Nancy Pelosi didn't get COVID-19 vaccine. The Facts: "Claims that Nancy Pelosi did not actually get vaccinated against COVID-19 because there was an orange cap over the needle are FALSE. Photos of Pelosi getting the vaccine clearly show the needle going into her arm without a cap on it." (Sadeghi 2020)

"A Facebook user has made multiple false claims about COVID-19 vaccines in a video that has been shared thousands of times. The user uploaded a video to Facebook on Jan. 25 in which he told his audience that having the vaccine "doesn't mean you are protected" and that its "most common" side-effects are anaphylactic shock and

Bell's palsy, adding that "you've got a 10-15% chance of catching one of those". He goes on to say that the ingredients will also make a patient more susceptible to catching HIV. His claim reads: "Just remember this when you've had your vaccine that they are telling us, number one, first of all, that having this vaccine doesn't mean you are protected. You can still catch COVID. Not only can you still catch COVID, coronavirus, you can still transmit it… You have a 15% chance of having an adverse reaction – so you've gone from a 0% chance from not having the vaccine and not having a cold or a sniffle to vaccinating yourself and giving yourself an uplift of 15% chance of having an adverse reaction. So the most common reaction so far to date are anaphylactic shock and bell's palsy I believe are the main side-effects that we're seeing. So, just to make you aware, you've got a 10-15% chance of catching one of those […] "Now on the ingredients… they say you've got a good chance of catching HIV from this vaccine, so it makes you more susceptible." The Facts: False. The vaccines for COVID-19 protect an individual by preventing serious outcomes of the disease. Scientists are still researching whether the vaccines also stop transmission. The most common symptoms of the vaccine include pain at the site of injection and redness – not anaphylaxis nor Bell's palsy. While four vaccine recipients in a US study developed Bell's palsy, it has not been connected to the vaccine. Researchers have also warned about susceptibility to HIV in vaccines using a specific viral vector, but no COVID-19 vaccines using this vector have been approved for use in the UK." (Reuters 2020) To add the transmission claim has been updated and we now know that due to a recent cohort study in Italy and the US protection against transmission was great news.

The claim: "Adm. Brett Giroir, an assistant secretary at the Department of Health and Human Services, said the COVID-19 vaccines have tracking mechanisms. The Facts: The claim that Brett Giroir, assistant secretary of Health and Human Services, said the COVID-19 vaccines have tracking mechanisms is MISSING CONTEXT, based on our research. Giroir said there were "tracking mechanisms" when discussing vaccine distribution, but he did not say they were in the vaccines themselves. Neither the Pfizer nor Moderna vaccine contains a tracking device. There are, however, systems that help vaccine providers and government officials track who has been vaccinated to ensure everyone receives a second dose." (Ngo 2021) Like other claims shared on social media context is always missing regarding claims centered on devices, tracking, and vaccines. This claim is a clear indication of that same behavior that continues to exist in the anti-vaccine arena.

The Claim: Lopinavir-Ritonavir can treat and cure Covid-19. The Facts: "A total of 199 patients with laboratory-confirmed SARS-CoV-2 infection underwent randomization; 99 were assigned to the lopinavir–ritonavir group, and 100 to the standard-care group. Treatment with lopinavir–ritonavir was not associated with a difference from standard care in the time to clinical improvement (hazard ratio for clinical improvement, 1.31; 95% confidence interval [CI], 0.95 to 1.80). Mortality at 28 days was similar in the lopinavir–ritonavir group and the standard-care group (19.2% vs. 25.0%; difference, −5.8 percentage points; 95% CI, −17.3 to 5.7). The percentages of patients with detectable viral RNA at various time points were similar. In a modified intention-to-treat analysis, lopinavir–ritonavir led to a median time to clinical improvement that was shorter by 1 day than that observed with standard care (hazard ratio, 1.39; 95% CI, 1.00 to 1.91). Gastrointestinal adverse events were more common in the lopinavir–ritonavir group, but serious adverse events were more common in the standard-care group. Lopinavir–ritonavir treatment was stopped early in 13 patients (13.8%) because of adverse events. CONCLUSIONS In hospitalized adult patients with severe Covid-19, no benefit was observed with lopinavir–ritonavir treatment beyond standard care. Future trials in patients with severe illness may help to confirm or exclude the possibility of a treatment benefit. (Funded by Major Projects of National Science and Technology on New Drug Creation and Development and others." (Bin Cao, M.D., Yeming Wang, M.D., Danning Wen, M.D. 2020)

"The claim: China recovered from the coronavirus without a vaccine. The Facts: Based on our research, the claim that China recovered from the coronavirus without a vaccine is PARTLY FALSE. China has succeeded in its efforts to control the spread of COVID-19 within its borders in large part because of the speed and strictness of its measures. But it has also used vaccines, and it actually began vaccinating large swaths of the population before the rest of the world. Just one company already has vaccinated 1 million citizens." (Caldera 2020)

As you can see from the above claims most are made based on ignorance, failure to fact check sources, and drawing false conclusions surrounding alternatives medicines. It's as if people make any excuse to fight against science. That has to change and I hope that this chapter provides a pathway of helping anyone who hear future claims about viruses and vaccines learn where the origin of the claims are coming from and how to fact check those claims and draw an honest conclusion before assuming the source to be accurate.

I owe Fact Checkers from USAToday, Reuters, Politico, Health Feedback, factchecker.org and Associated Press and huge thank you. Because each source provided additional sources that allowed me to cross reference the claims and the facts. Learning how to use Research Methodology has influenced my ability to track down claims and verify them. I spent the entire 2020 year refuting claims on social media. What I did not find from each fact checking site I looked to open access articles published on alternative medicinal trials for data. I would also like to thank New England Journal of Medicine, Nature, Science Mag, Jama Network, The Lancet, Medrxiv, Science Direct, and a few others for opening access to the public and allow those of us who have paid attention to learn from the experts and understand the science relating to Covid-19 and 3rd Generation Vaccines. This chapter is the result of What Sars-CoV2 Taught Me and I hope you learned from this chapter specifically and how to take your time to research and verify claims made by those on social media.

Trust the Science not the Scientist/or Claimant

Source up or Shut up!

Chapter VI Sources

- Affairs, Office of Regulatory. "Fraudulent COVID-19 Products." U.S. Food and Drug Administration, FDA, 10 June 2021, www.fda.gov/consumers/health-fraud-scams/fraudulent-coronavirus-disease-2019-covid-19-products.

- Ahmed, Irman. "The Disinformation Dozen: Center for Countering Digital Hate." CCDH, CCDH , 24 Mar. 2021, www.counterhate.com/disinformationdozen.

- Bill McCarthy, PolitiFact.com. "Fact-Check: Do COVID Vaccines 'Magnetize' People?" Statesman, Austin American-Statesman, 11 June 2021, www.statesman.com/story/news/politics/politifact/2021/06/11/sherri-tenpenny-makes-false-covid-vaccine-magnetism-claim-lawmakers/7653766002/.

- Caldera, Camille. "Fact Check: Operation Warp Speed Official Discussed Vaccine Distribution, Not Mandatory Vaccinations." USA Today, Gannett Satellite Information Network, 25 Nov. 2020, www.usatoday.com/story/news/factcheck/2020/11/24/fact-check-post-operation-warp-speed-official-missing-context/6398580002/.

- Caldera, Camille. "Fact Check: Strict Lockdowns, Experimental Vaccine Helped China Recover from COVID-19." USA Today, Gannett Satellite Information Network, 8 Dec. 2020, www.usatoday.com/story/news/factcheck/2020/12/04/fact-check-strict-lockdowns-covid-19-vaccine-helped-china-recover/3814394001/.

- Cao, Bin, et al. "A Trial of Lopinavir–Ritonavir in Adults Hospitalized with Severe Covid-19: NEJM." New England Journal of

Medicine, 7 May 2020, www.nejm.org/doi/full/10.1056/NEJMoa2001282.

- Carballo-Carbajal, Iria. "No, Bill Gates Is Not Funding COVID-19 Vaccines as a Way to Conduct Global Surveillance or to Depopulate the World." Health Feedback, 8 July 2020, healthfeedback.org/claimreview/no-bill-gates-is-not-funding-covid-19-vaccines-as-a-way-to-conduct-global-surveillance-or-to-depopulate-the-world/.

- CC, MSK. "Fact Check: 7 Persistent Myths about COVID-19 Vaccines." Memorial Sloan Kettering Cancer Center, 24 May 2021, www.mskcc.org/coronavirus/myths-about-covid-19-vaccines.

- Clinic Staff, Mayo. "Vasovagal Syncope." Mayo Clinic, Mayo Foundation for Medical Education and Research, 19 Feb. 2021, www.mayoclinic.org/diseases-conditions/vasovagal-syncope/symptoms-causes/syc-20350527.

- Crist, Carolyn. "'Disinformation Dozen' Driving Anti-Vaccine Content." WebMD, WebMD, 25 Mar. 2021, www.webmd.com/children/vaccines/news/20210325/disinformation-dozen-driving-anti-vaccine-content.

- Dupuy, Beatrice. "Experts: MRNA Vaccine for COVID-19 Does Not Alter DNA." AP NEWS, Associated Press, 4 Sept. 2020, apnews.com/article/archive-fact-checking-9340521654.

- Fack Checker, Reuters. "Fact Check-Moderna's Chief Medical Officer Did Not Say MRNA Vaccines Alter DNA." Reuters, Thomson Reuters, 8 Apr. 2021, www.reuters.com/article/factcheck-moderna-mrna/fact-check-modernas-chief-medical-officer-did-not-say-mrna-vaccines-alter-dna-idUSL1N2M10IV.

- Fack Checkers, Reuters. "Fact Check: Video Makes Multiple False Claims about COVID-19 Pandemic." Reuters, Thomson Reuters, 3 Nov. 2020, www.reuters.com/article/uk-factcheck-pandemic-video/fact-check-video-makes-multiple-false-claims-about-covid-19-pandemic-idUSKBN27J2HM.

- Fact Check, Reuters. "Fact Check-No Evidence MRNA COVID-19 Vaccines Affect Sperm." Reuters, Thomson Reuters, 17 May 2021, www.reuters.com/article/factcheck-sperm-vaccine/fact-check-no-evidence-mrna-covid-19-vaccines-affect-sperm-idUSL2N2N42EC.

- Fact Checker, Reuters. "Fact Check: Anaphylaxis and Bell's Palsy Are Not the Most Common Side-Effects of COVID-19 Vaccines." Reuters, Thomson Reuters, 30 Jan. 2021, www.reuters.com/article/uk-factcheck-anaphylaxis/fact-check-anaphylaxis-and-bells-palsy-are-not-the-most-common-side-effects-of-covid-19-vaccines-idUSKBN29Z0PD.

- Fact Checker, Reuters. "Fact Check: Lipid Nanoparticles in a COVID-19 Vaccine Are There to Transport RNA Molecules." Reuters, Thomson Reuters, 5 Dec. 2020, www.reuters.com/article/uk-factcheck-vaccine-nanoparticles/fact-check-lipid-nanoparticles-in-a-covid-19-vaccine-are-there-to-transport-rna-molecules-idUSKBN28F0I9.

- Fact Checker, Reuters. "Partly False Claim: A List of Eight Coronavirus-Related 'Facts.'" Reuters, Thomson Reuters, 26 Mar. 2020, www.reuters.com/article/uk-factcheck-coronavirus-eight-facts/partly-false-claim-a-list-of-eight-coronavirus-related-facts-idUSKBN21D3EY.

- Funke, Daniel. "PolitiFact - Alternative Health Website Spreads False Claim about COVID-19 Vaccine Side Effects." @Politifact, 9

Dec. 2020, www.politifact.com/factchecks/2020/dec/09/blog-posting/alternative-health-website-spreads-false-claim-abo/.

- Gore, D'Angelo. "Hank Aaron's Death Attributed to Natural Causes." FactCheck.org, 28 Apr. 2021, www.factcheck.org/2021/01/scicheck-hank-aarons-death-attributed-to-natural-causes/.

- Greenberg, Jon. "PolitiFact - COVID-19 Skeptics Say There's an Overcount. Doctors in the Field Say the Opposite." @Politifact, 14 Apr. 2020, www.politifact.com/factchecks/2020/apr/14/candace-owens/covid-19-skeptics-say-theres-overcount-doctors-fie/.

- Gruber-Miller, Stephen. "Fact Check: Vitamins C and D Are Not Used in 'Conventional Treatment' of Coronavirus." USA Today, Gannett Satellite Information Network, 4 May 2020, www.usatoday.com/story/news/factcheck/2020/05/02/fact-check-coronavirus-covid-19-vitamins-c-d-treatment-joseph-mercola/3058491001/.

- Hui, Kayla. "CDC Study Confirms That COVID-19 Vaccines Block Transmission In the Real World." Verywell Health, 8 Apr. 2021, www.verywellhealth.com/cdc-study-covid-19-transmission-vaccines-5121080.

- Jones, Craig. "Claim That the FDA Found That Coronavirus Vaccines Awaiting Approval Could Cause Death Is Majorly Misleading." Newswise, 10 Dec. 2020, www.newswise.com/factcheck/claim-that-the-fda-found-that-coronavirus-vaccines-awaiting-approval-could-cause-death-is-majorly-misleading.

- Kancharla, Bharath, and Nanditha Kalidoss. "The Claims Made by Dr. Christiane Northrup Regarding COVID-19 Vaccines in This Video Are False." FACTLY, Factly, 20 Nov. 2020, factly.in/the-claims-made-by-dr-christiane-northrup-regarding-covid-19-vaccines-in-this-video-are-false/.

- Kiley, James P. "NIH Halts Clinical Trial of Hydroxychloroquine." National Institutes of Health, U.S. Department of Health and Human Services, 20 June 2020, www.nih.gov/news-events/news-releases/nih-halts-clinical-trial-hydroxychloroquine.

- Lovelace Jr, Berkeley, and Kevin Breuninger . "Trump Says He Takes Hydroxychloroquine to Prevent Coronavirus Infection Even Though It's an Unproven Treatment." CNBC, CNBC, 19 May 2020, www.cnbc.com/2020/05/18/trump-says-he-takes-hydroxychloroquine-to-prevent-coronavirus-infection.html.

- Mason, Jacquelyn. "The Nation of Islam and Anti-Vaccine Rhetoric." First Draft, 27 May 2021, firstdraftnews.org/articles/the-nation-of-islam-and-anti-vaccine-rhetoric/.

- McCarthy, Bill, et al. "Sherri Tenpenny." PolitiFact, 2021, www.politifact.com/personalities/sherri-tenpenny/.

- McLernon , Lianna Matt. "Zinc, Vitamin C Show No Effect for COVID-19 in Small Study." CIDRAP, 12 Feb. 2021, www.cidrap.umn.edu/news-perspective/2021/02/zinc-vitamin-c-show-no-effect-covid-19-small-study.

- Murai, Igor H, et al. "Effect of a Single High Dose of Vitamin D3 in Patients With Moderate to Severe COVID-19." JAMA, JAMA Network, 16 Mar. 2021, jamanetwork.com/journals/jama/fullarticle/2776738.

- Ngo, Madeleine. "Fact Check: Health and Human Services' Brett Giroir Confirms Vaccine Distribution Is Tracked to Ensure Dosing." USA Today, Gannett Satellite Information Network, 19 Jan. 2021, www.usatoday.com/story/news/factcheck/

2021/01/19/fact-check-covid-19-vaccines-pfizer-moderna-brett-giroir-tracking/4207433001/.

- Pitts, D'Angela. "Here's Where That COVID-19 Vaccine Infertility Myth Came From-And Why It Is Not True." Henry Ford LiveWell, Henry Ford Health System Staff, 23 Apr. 2021, www.henryford.com/blog/2021/04/fertility-rumor-covid-vaccine.

- Press, Associated. "Reno Doctor's Selfie Used to Claim COVID-19 Is a Hoax | Health News | US News." U.S. News & World Report, U.S. News & World Report, 1 Dec. 2020, www.usnews.com/news/health-news/articles/2020-12-01/reno-doctors-selfie-used-to-claim-covid-19-is-a-hoax.

- Press, The Associated. "Post Makes False Claim about COVID-19 Vaccine Risk." AP NEWS, Associated Press, 1 Feb. 2021, apnews.com/article/fact-checking-afs:Content:9934822788.

- Press, The Associated. "Research Shows COVID-19 Was Not Manufactured in a Lab." AP NEWS, Associated Press, 16 Sept. 2020, apnews.com/article/archive-fact-checking-9391149002.

- Rougerie, Pablo. "No Aborted Fetal Tissue or Cells in the Johnson & Johnson COVID-19 Vaccine." Health Feedback, 25 Mar. 2021, healthfeedback.org/claimreview/no-aborted-fetal-tissue-or-cells-in-the-johnson-johnson-covid-19-vaccine/.

- Rourke, Ciara. "PolitiFact - People Are Dying from COVID-19." @Politifact, 4 Sept. 2020, www.politifact.com/factchecks/2020/sep/04/viral-image/people-are-dying-covid-19/.

- Sadeghi, McKenzie. "Fact Check: Posts Falsely Claim Cap Was on Nancy Pelosi's COVID-19 Vaccine." USA Today, Gannett Satellite Information Network, 19 Dec. 2020, www.usatoday.com/story/news/factcheck/2020/12/19/fact-check-photos-show-nancy-pelosi-receiving-covid-19-vaccine/3973360001/.

- Self, Wesley H, et al. "Effect of Hydroxychloroquine on Clinical Status at 14 Days in Hospitalized Patients With COVID-19." JAMA, JAMA Network, 1 Dec. 2020, jamanetwork.com/journals/jama/fullarticle/2772922.

- Skeptic, Original. "Kelly Brogan Archives." Skeptical Raptor, 2 May 2021, www.skepticalraptor.com/skepticalraptorblog.php/tag/kelly-brogan/.

- Smith, Michelle R, and Johnatan Reiss. "How Vaccine Disinformation Super Spreaders Have Cashed in on Americans' Fears during Pandemic." Chicagotribune.com, 13 May 2021, www.chicagotribune.com/coronavirus/vaccine/ct-aud-nw-vaccine-disinformation-20210513-7lid6y5jlzakdjow3wfzwwxruy-story.html.

- Spencer , Saranac Hale, and Angelo Fichera. "RFK Jr. Video Pushes Known Vaccine Misrepresentations." FactCheck.org, 28 Apr. 2021, www.factcheck.org/2021/03/scicheck-rfk-jr-video-pushes-known-vaccine-misrepresentations/.

- Staff, Reuters. "Fact Check: RFID Microchips Will Not Be Injected with the COVID-19 Vaccine, Altered Video Features Bill and Melinda Gates and Jack Ma." Reuters, Thomson Reuters, 4 Dec. 2020, www.reuters.com/article/uk-factcheck-vaccine-microchip-gates-ma/fact-check-rfid-microchips-will-not-be-injected-with-the-covid-19-vaccine-altered-video-features-bill-and-melinda-gates-and-jack-ma-idUSKBN28E286.

- Strauss, Valerie. "Analysis | Debunking Anti-Vaxxer RFK Jr.'s Claim about 'Suspicious' Coronavirus Vaccine Deaths, a Phony Elon Musk Tweet and More News Literacy Lessons." The Washington Post, WP Company, 5 Feb. 2021, www.washingtonpost.com/education/2021/02/05/news-literacy-refuting-rfkjr-phony-elon-musk-tweet/.

- Tech, Fora. "In Spite of Evidence to the Contrary, Rizza Islam Claims That the MMR Vaccine Is Used for Depopulation." Science Feedback, Health Feedback, 19 Dec. 2019, sciencefeedback.co/claimreview/in-spite-of-evidence-to-the-contrary-rizza-islam-claims-that-the-mmr-vaccine-is-used-for-depopulation/.

- Teoh, Flora. "Frequency of Deaths in Elderly Individuals after COVID-19 Vaccination Wasn't Higher than the Frequency in Those Who Weren't Vaccinated." Health Feedback, 27 Jan. 2021, healthfeedback.org/claimreview/frequency-of-deaths-in-elderly-individuals-after-covid-19-vaccination-werent-higher-than-the-frequency-in-those-who-werent-vaccinated/.

- Teoh, Flora. "The U.S. National Childhood Vaccine Injury Act Does Not Stop People from Suing Vaccine Manufacturers." Health Feedback, 7 Apr. 2021, healthfeedback.org/claimreview/the-u-s-national-childhood-vaccine-injury-act-does-not-stop-people-from-suing-vaccine-manufacturers/.

- Thomas, Suma, et al. "Effect of Zinc and Ascorbic Acid on Symptom Length Among Patients With SARS-CoV-2." JAMA Network Open, JAMA Network, 12 Feb. 2021, jamanetwork.com/journals/jamanetworkopen/fullarticle/2776305.

- UC Davis Health, Public Affairs and Marketing. "UC Davis Experts: Science Says Wearing Masks and Social Distancing Slow COVID-19 (VIDEO)." UC Davis Health, 6 July 2020, health.ucdavis.edu/health-news/newsroom/uc-davis-experts-science-says-wearing-masks-and-social-distancing-slow-covid-19/2020/07.

- USA, EPA. "Protect Your Family from Sources of Lead." EPA, Environmental Protection Agency, 22 Dec. 2020, www.epa.gov/lead/protect-your-family-sources-lead#:~:text=Deteriorating%20lead%2Dbased%20paint%20(peeling,Doors%20and%20door%20frames%3B%20and.

- VĒBERE, ILZE. "Diskreditēts Osteopāts Izplata Melus Par Covid-19 Pandēmiju." A Discredited Osteopath Spreads Lies about the Covid-19 Pandemic, 26 May 2020, rebaltica.lv/2020/05/vai-tiesam-viens-cilveks-fauci-vainojams-covid-19-pandemija/.

- Zubeyir, Ayşe E. "Kendini Doktor Olarak Tanıtan Ben Tapper'ın PCR Testiyle Ilgili Iddiaları: Teyit." Edited by Sosyal M Medya, Şüpheli Bilgileri Inceleyen Doğrulama Platformu, Teyit, 22 Dec. 2020, teyit.org/analiz-dr-ben-tapperin-pcr-testine-yonelik-iddialari.

Sars-CoV2 Origins
Bonus Material

The Origins of Covid are still being investigated; however, a very shaky hypothesis has been regurgitated by nonscientists stating that the virus was man-made or potentially leaked from a lab. In chapter 1, I began with viruses for a reason because I needed to make sure that you understand viruses are natural. As we began to populate the globe, the more we became exposed to novel viruses. Currently, cell.com has published a nice article on Sars-CoV2 origins in bats. Results demonstrated a 96% - 97% match providing a very strong case. However, we will continue to allow science an opportunity to have its say and not come to an absolute conclusion at this time. WHO (World Health Organization) was not convinced of the lab leaked origins but did not close the door on that option. The Who study revealed two potential lineages of concern. The wet market in Wuhan and domesticated wildlife of some form. WHO stated tracebacks are very important in this stage.

As for virus origins, if you think The Spanish Flu evolved from Spain, that shows a lack of understanding of viruses on your part. Its origins have not been settled for over 100 years. Science has no consensus; however, historians assume the origins are in China. What many do not know about virus origins is that it is hard to narrow down a virus origin to a specific place, but after further research, we can potentially narrow it down to a geographical location. Swine flu was assumed to have its origins in the U.S., but later it was revealed that its true origins were Mexico.

"There is much debate among virologists about this question. Three main hypotheses have been articulated: 1. The progressive, or escape, hypothesis states that viruses arose from genetic elements that gained the ability to move between cells; 2. The regressive or reduction hypothesis asserts that viruses are remnants of cellular organisms; and 3. The virus-first hypothesis states that viruses predate or coevolved with their current cellular hosts. The Progressive Hypothesis: According to this hypothesis, viruses originated through a progressive process. Mobile genetic elements, pieces of genetic material capable of moving within a genome, gained the ability to exit one cell and enter another. To conceptualize this transformation, let's examine the replication of retroviruses, the family of viruses to which HIV belongs. Retroviruses have a single stranded RNA genome.

When the virus enters a host cell, a viral enzyme, reverse transcriptase, converts that single-stranded RNA into double-stranded DNA. This viral DNA then migrates to the nucleus of the host cell. Another viral enzyme, integrase, insert the newly formed viral DNA into the host cell's genome. Viral genes can then be transcribed and translated. The host cell's RNA polymerase can produce new copies of the virus's single stranded RNA genome. Progeny viruses assemble and exit the cell to begin the process again. This process very closely mirrors the movement of an important, though somewhat unusual, component of most eukaryotic genomes: retrotransposons. These mobile genetic elements make up an astonishing 42% of the human genome (Lander et al. 2001) and can move within the genome via an RNA intermediate. Like retroviruses, certain classes of retrotransposons, the viral-like retrotransposons, encode a reverse transcriptase and, often, an intergrase. With these enzymes, these elements can be transcribed into RNA, reverse-transcribed into DNA, and then integrated into a new location within the genome. We can speculate that the acquisition of a few structural proteins could allow the element to exit a cell and enter a new cell, thereby becoming an infectious agent. Indeed, the genetic structures of retroviruses and viral-like retrotransposons show remarkable similarities. The Regressive Hypothesis: In contrast to the progressive process just described, viruses may have originated via a regressive or reductive process. Microbiologists generally agree that certain bacteria that are obligate intracellular parasites, like Chlamydia and Rickettsia species, evolved from free-living ancestors. Indeed, genomic studies indicate that the mitochondria of eukaryotic cells and Rickettsia prowazekii may share a common, free-living ancestor (Andersson et al. 1998). It follows, then, that existing viruses may have evolved from more complex, possibly free-living organisms that lost genetic information over time as they adopted a parasitic approach to replication. Viruses of one particular group, the nucleocytoplasmic large DNA viruses (NCLDVs), best illustrate this hypothesis. These viruses, which include smallpox virus and the recently discovered giant of all viruses, Mimivirus, are much bigger than most viruses (LaScola et al., 2003). A typical brick-shaped poxvirus, for instance, may be 200 nm wide and 300 nm long. About twice that size, Mimivirus exhibits a total diameter of roughly 750 nm (Xiao et al. 2005). Conversely, spherically shaped influenza virus particles maybe only 80 nm in diameter, and poliovirus particles have a diameter of only 30 nm, roughly 10,000 times smaller than a grain of salt. The NCLDVs also possess large genomes. Again, poxvirus genomes often approach 200,000 base pairs, and Mimivirus has a genome of 1.2 million base pairs, while poliovirus has a genome of only 7,500 nucleotides total. In addition to their large size, the NCLDVs exhibit greater complexity than other viruses have and depend less on their host for replication than do other viruses. Poxvirus particles, for instance, include a large number of viral enzymes and related factors that allow the virus to produce functional messenger RNA within the host cell cytoplasm. Because of the size and complexity of NCLDVs, some virologists have hypothesized that these viruses may be descendants of more complex ancestors. According to proponents of this hypothesis, autonomous organisms initially developed a symbiotic relationship. Over time, the relationship turned parasitic, as one organism became more and more dependent on the other. As the once free-living parasite became more dependent on the host, it lost previously essential genes. Eventually, it was unable to replicate independently, becoming an obligate intracellular parasite, a virus. Analysis of the giant Mimivirus may support this hypothesis. This virus contains a relatively large repertoire of putative genes associated with translation — genes that may be remnants of a previously complete translation system. Interestingly, Mimivirus does not differ appreciably from parasitic bacteria, such as Rickettsia prowazekii (Raoult et al. 2004). The Virus-First Hypothesis: The progressive and regressive hypotheses both assume that cells existed before viruses. What if viruses existed first? Recently, several investigators proposed that viruses may have been the first replicating entities. Koonin and Martin (2005) postulated that viruses existed in a precellular world as self replicating units. Over time these units, they argue, became more organized and more complex. Eventually, enzymes for the synthesis of membranes and cell walls evolved, resulting in the formation of cells. Viruses, then, may have existed before bacteria, archaea, or eukaryotes

(Prangishvili et al. 2006). Most biologists now agree that the very first replicating molecules consisted of RNA, not DNA. We also know that some RNA molecules, ribozymes, exhibit enzymatic properties; they can catalyze chemical reactions. Perhaps, simple replicating RNA molecules, existing before the first cell formed, developed the ability to infect the first cells. Could today's single-stranded RNA viruses be descendants of these precellular RNA molecules? Others have argued that precursors of today's NCLDVs led to the emergence of eukaryotic cells. Villarreal and DeFilippis (2000) and Bell (2001) described models explaining this proposal. Perhaps, both groups postulate, the current nucleus in eukaryotic cells arose from an endosymbiotic-like event in which a complex, enveloped DNA virus became a permanent resident of an emerging eukaryotic cell. No Single Hypothesis May Be Correct: Where viruses came from is not a simple question to answer. One can argue quite convincingly that certain viruses, such as the retroviruses, arose through a progressive process. Mobile genetic elements gained the ability to travel between cells, becoming infectious agents. One can also argue that large DNA viruses arose through a regressive process whereby once-independent entities lost key genes over time and adopted a parasitic replication strategy. Finally, the idea that viruses gave rise to life as we know it presents very intriguing possibilities. Perhaps today's viruses arose multiple times, via multiple mechanisms. Perhaps all viruses arose via a mechanism yet to be uncovered. Today's basic research in fields like microbiology, genomics, and structural biology may provide us with answers to this basic question." (Wessner, Ph.D.: 2010)

This is important to know when researching virus origins; one can't succumb to lab leak hypothesis because without understanding we cant determine how it got in a lab, to begin with? Scientists strongly concluded that the virus has not been modified. "More specifically, the authors of the new research looked at two components of spike proteins: the receptor-binding domain (RBD), which latches onto healthy host cells, and the cleavage site, which opens up the virus and allows it to penetrate the host cell. To bind to human cells, spike proteins need a receptor on human cells called angiotensin-converting enzyme 2 (ACE2). The scientists found that the receptor-binding domain of the spike protein had evolved to target ACE2 so effectively that it could only have been the result of natural selection and not of genetic engineering. Furthermore, the molecular structure of the backbone of SARS-CoV-2 supported this finding. If scientists had engineered the new coronavirus purposely as a pathogen, explain the researchers, the starting point would likely have been the backbone of another virus in the coronavirus family. However, the backbone of SARS-CoV-2 was very different than those of other coronaviruses and was most similar to related viruses in bats and pangolins. "These two features of the virus — the mutations in the RBD portion of the spike protein and its distinct backbone — rule out laboratory manipulation as a potential origin for SARS-CoV-2," explains Andersen." (Sandoiu 2020)

Conspiracists claim without any evidence that people have modified and created the (ACE2), citing gain of function as a premise. When asked to explain gain of function, things began to become murky, which is a sign that someone is regurgitating something someone else had said without providing any evidence.

As mentioned previously, the study published in Cell uncovered 4 related novel coronaviruses found in bats. More specifically, rhinolophid bats which were found in several different parts of Asia. "Despite the discovery of animal coronaviruses related to SARS-CoV-2, the evolutionary origins of this virus are elusive. We describe a meta-transcriptomic study of 411 bat samples collected from a small geographical region in Yunnan province, China, between May 2019 and November 2020. We identified 24 full length coronavirus genomes, including four novel SARS-CoV-2-related and three SARS-CoV-related viruses. Rhinolophus pusillus virus RpYN06 was the closest relative of SARS-CoV-2 in most of the genome, although it possessed a more divergent spike gene. The other three SARS-CoV-2-related coronaviruses carried a genetically distinct spike gene that could weakly bind to the hACE2 receptor in vitro. Ecological modeling predicted the coexistence of up to 23 Rhinolophus bat species, with the largest contiguous hotspots extending from South

Laos and Vietnam to southern China. Our study highlights the remarkable diversity of bat coronaviruses at the local scale, including close relatives of both SARS-CoV-2 and SARS-CoV." (Zhou, Ji, Chen, Wang, Hu 2021)

"The genetic lineages of SARS-CoV-2 associated with early cases in Wuhan were documented in the WHO report. Previously, Rambaut et al, (2020) noted that at the root of the phylogeny of SARS-CoV-2 are two lineages designated lineage A and B. Early lineage A viruses include SARS-Cov-2 isolate EPI_ISL_529213 sampled on 30Dec19 from a person linked to a wildlife market different than the Huanan market (Molecular Epidemiology, sample 13 in WHO, 2021). Linage A viruses share two nucleotides (T8,782 in ORF1ab and C28,144 in ORF8) with the bat viruses RaTG13 and RmYN02 and other sarbecoviruses. It is likely that the most recent common ancestor (MRCA) of SARS-CoV-2 shares the same genome sequence as these early lineage A sequences (Rambaut et al., 2020). Different nucleotides (C8,782 in ORF1ab and T28,144 in ORF8) are present at those sites in viruses assigned to lineage B, such as SARS-CoV-2 isolate Wuhan-Hu-1 (GenBank accession no. MN908947, Molecular Epidemiology, sample 06 in WHO, 2021) sampled from the Huanan market on 30Dec19. All virus positive samples from the Huanan market, from venders or customers of the market or from environmental samples, contained SARS-CoV-2 of lineage B. There was limited genetic diversity in the lineage B samples from the Huanan market, which is consistent with the market as a site of a super-spreader event. Lineage A and Lineage B viruses spread throughout Wuhan and to other countries" (Rambaut et al., 2020; Worobey et al., 2020). (Garry 2021) In short lineage A and B are seen flowing toward the area not coming from the opposite direction. This is a key, because it adds to all the evidence that has been compiled over the past year and with reports that have surfaced online. While writing this book we have also learned that the White House intel was inconclusive and did not lean one way or another.

Until the science is settled, I suggest people focus on staying safe, continue to practice social distancing, wearing masks, and get vaccinated. This is our best line of defense against a novel virus with multiple variants. However, once we understand virus origins and each hypothesis, we can conclude that we may not ever really get an exact answer of Sars CoV2 origins. What Cell published is a clear indicator that Sars-Cov2 was not made in a lab and existed in nature. The lab leak theory is just a hypothesis, not a theory because that claim has not been scrutinized scientifically, nor has it gone through rigorous studies to demonstrate it to be plausible otherwise.

And…….. This is What Sars CoV2 Taught Me!

Bonus (Summary) Sources

- Zhou, Hong, et al. "Identification of Novel Bat Coronaviruses Sheds Light on the Evolutionary Origins of SARS-CoV-2 and Related Viruses." Cell, Cell Press, 9 June 2021, www.sciencedirect.com/science/article/pii/S0092867421007091?via%3Dihub.

- Andersen, Kristian G., et al. "The Proximal Origin of SARS-CoV-2." Nature News, Nature Publishing Group, 17 Mar. 2020, www.nature.com/articles/s41591-020-0820-9#Sec8.

- Sandoiu, Ana. "The New Coronavirus Was Not Man-Made, Study Shows." Medical News Today, MediLexicon International, 20 Mar. 2020, www.medicalnewstoday.com/articles/the-new-coronavirus-was-not-genetically-engineered-study-shows#Ending-the-rumors-about-SARS-CoV-2-.

- Wessner, David R. Nature News, Nature Publishing Group, 2010, www.nature.com/scitable/topicpage/the-origins-of-viruses-14398218/.

- Farry, R F, et al. "Early Appearance of Two Distinct GENOMIC Lineages OF SARS-CoV-2 in Different Wuhan Wildlife Markets Suggests SARS-CoV-2 Has a Natural Origin." *Virological*, 12 May 2021, virological.org/t/early-appearance-of-two-distinct-genomic-lineages-of-sars-cov-2-in-different-wuhan-wildlife-markets-suggests-sars-cov-2-has-a-natural-origin/691.

Understanding Breakthrough Infections & Vaccine Effectiveness

Bonus Material

"The term 'breakthrough infection' is wrong. Most human vaccines do not prevent infection. They prevent disease, which is what Covid-19 vaccines were tested for. It doesn't matter if infections occur; what matters is moderate to serious disease" - (Vincent Racaniello 2021). As breakthrough infections emerge across the country anti-vaxxers made it a priority to focus on vaccines not being affective at preventing infection as if viruses somehow dissipate. In the middle of a pandemic, vaccine hesitancy in the public has been on the rise, and most of it stems from a lack of vaccine understanding. Regardless of how much effort scientists, doctors, and nurses have utilized to properly educate the public, misinformation seems to spread a lot faster than the truth. My goal is to elaborate on what Vincent Racaniello stated but also make sure that we have at least the basic understanding of what vaccines are if they are failing and if vaccinated people are transmitting the virus.

What are vaccines? "Vaccines are products that protect people against many diseases that can be very dangerous and even deadly. Different from most medicines that treat or cure diseases, vaccines prevent you from getting sick with the disease in the first place" - (ImmunizeBC 2021). To add, no vaccine is 100% effective at preventing disease, and we know this to be certain because of our history of immunization in the New World. The initial vaccine efficacy prior to mRNA vaccine data required by the FDA was set at 50% - 60%. Scientists urged the FDA that efficacy rates should be just a tad bit higher, somewhere around 70% - 75%. Flu vaccine efficacy is around the initial rate mRNA vaccines were required, but as the year progress and data became available, both Pfizer and Moderna vaccines proved to be 94% - 95%, respectively against preventing disease. "Vaccines prevent diseases that can be dangerous, or even deadly. Vaccines greatly reduce the risk of infection by working with the body's natural defenses to safely develop immunity to disease." (CDC 2021)

Efficacy is determined by how well a vaccine prevents the spread of disease. This is important to know because without this data, most vaccines would not be approved. But how it is measured requires proper understanding, and according to WHO: "A vaccine's efficacy is measured in a controlled clinical trial and is based on how many people who got vaccinated developed the 'outcome of

interest' (usually disease) compared with how many people who got the placebo (dummy vaccine) developed the same outcome. Once the study is complete, the numbers of sick people in each group are compared in order to calculate the relative risk of getting sick depending on whether or not the subjects received the vaccine. From this, we get the efficacy – a measure of how much the vaccine lowered the risk of getting sick. If a vaccine has high efficacy, a lot fewer people in the group who received the vaccine got sick than the people in the group who received the placebo. So, for example, let's imagine a vaccine with a proven efficacy of 80%. This means that – out of the people in the clinical trial – those who received the vaccine were at a 80% lower risk of developing disease than the group who received the placebo. This is calculated by comparing the number of cases of disease in the vaccinated group versus the placebo group. An efficacy of 80% does not mean that 20% of the vaccinated group will become ill." (WHO 2021)

Knowing a vaccines efficacy allows us to determine in a real-world setting potentially just how many people who be at risk for catching a disease, however, social media would have you think that the vaccines are not preventing disease. A recent headline made news rounds regarding the Delta variants comparison to chickenpox. It read, "Breaking News: The Delta variant is as contagious as chickenpox and may be spread by vaccinated people as easily as the unvaccinated, an internal C.D.C. report said." (New York Times 2020) This headline was shared over ten thousand times on Twitter, probably over one hundred thousand times between Facebook and Instagram, and we can't forget about TikTok. What many missed in that headline is the words "maybe" and inside the full news article it re-emphasized that, yet the headline was used to create even more doubt among people who are vaccine-hesitant and anti-vaxxers.

Many people are arguing regarding the headline that vaccinated people can spread the virus, and that is the narrative that our major news networks and social media sites have projected. Anti-vaxxers have claimed that vaccines are failing siting the CDC claiming they stated that vaccinated people are super-spreaders. Dr. Rochelle Walensky, director of the U.S. Centers for Disease Control and Prevention, said that vaccinated people were not super spreaders and the small outbreak in Massachusetts proved that the vaccines are working because it led to few hospitalizations, no deaths, and demonstrated vaccines are working against COVID-19. Also, vaccinated people are not responsible for the mutation of new variants. Since the onset of the pandemic, nextstrain.com has been monitoring and tracking variants and their mutations. When viruses adapt and find a new host (unvaccinated people), they began to mutate and spread, either creating a weaker version of itself or a more transmissible and infectious version of the virus. As we have witnessed since the start of the pandemic, as the virus spread from one continent to another, one version of the virus possessed something the other version did not. However, in a recent study published on Aug. 6th it revealed that unvaccinated people were 2.34 times more likely to contract Covid-19 again. This study, also published in the CDC website, did not only settle a long-standing debate but re-emphasize the importance of being vaccinated. While anti-vaxxers focused on making headlines, they failed to acknowledge that unvaccinated people are even more susceptible to reinfection and that reinfection can be severe.

So what is the deal with breakthrough infections I'll tell you…. It is completely blown out of proportion. According to the American Medical Association, "Nevertheless, the fact remains that getting vaccinated is effective in preventing people from getting severely ill or dying from the disease. Even as new COVID-19 variants appear, vaccines continue to hold their ground. But since no vaccine is perfect, it is expected that we will see some COVID-19 breakthrough infections. More than 161 million people in the U.S. have been fully vaccinated against COVID-19, according to the Centers for Disease Control and Prevention (CDC). Meanwhile, the CDC reports that there have been more than 10,000 COVID-19 vaccine breakthrough cases.

However, while the CDC initially tracked all breakthrough COVID-19 infections, as of May 1 the agency shifted to only tracking those linked to hospitalization or death. Over 5,100 patients with COVID-19 vaccine breakthrough infections were hospitalized or died. To add to that, 25 states report some data on COVID-19 breakthrough events, with 15 of those states regularly updating their data, according to Kaiser Family Foundation data. Data from these states indicate that breakthrough COVID-19 cases, hospitalizations, and deaths are extremely rare among those who are fully vaccinated, with the rate of infection well below 1% in all reporting states. The agency defines a breakthrough COVID-19 infection as "a small percentage of fully vaccinated persons" who "will still get COVID-19 if they are exposed to the virus that causes it." To that end, these vaccine breakthrough cases mean that "while people who have been vaccinated are much less likely to get sick, it will still happen in some cases." (Berg 2021)

Experts expected breakthrough infections, and to be honest, with vaccine efficacy being around 5% - 6% of 166 million people vaccinated. We have yet to see millions of reported breakthrough infections, but what has been reported since January has been 99% of deaths are in unvaccinated people while 98% of unvaccinated people have been hospitalized. "Less than 1 percent (about 0.004 percent) of fully vaccinated individuals have been hospitalized with, or have died from, COVID-19, according to the latest data tracked by the Centers for Disease Control and Prevention (CDC). When it comes to older adults, who were hit hardest by the coronavirus during its initial sweep, fully vaccinated individuals ages 65 and up are 94 percent less likely to be hospitalized with COVID-19 than people of the same age who are not vaccinated." (Nania 2021)

"People who receive mRNA COVID-19 vaccines are up to 91 percent less likely to develop the disease than those who are unvaccinated, according to a new nationwide study of eight sites, including Salt Lake City. For those few vaccinated people who do still get an infection or "breakthrough" cases, the study suggests that vaccines reduce the severity of COVID-19 symptoms and shorten its duration. Researchers say these results are among the first to show that mRNA vaccination benefits even those individuals who experience breakthrough infections. "One of the unique things about this study is that it measured the secondary benefits of the vaccine," says Sarang Yoon, D.O., a study co-author, assistant professor at the University of Utah Rocky Mountain Center for Occupational and Environmental Health (RMCOEH). Principal investigator of the RECOVER (Research on the Epidemiology of SARS-CoV-2 in Essential Response Personnel) study in Utah. The study, published online in the New England Journal of Medicine, builds on preliminary data released by the Centers for Disease Control and Prevention (CDC) in March. The study was designed to measure the risks and rates of infection among those on the front lines of the pandemic." (Dollemore 2021) As Vincent Racaniello stated at the term breakthrough infection is wrong, and we are confirming that with the data to support the position of one of America's leading virologists takes.

In closing, it is best to let science settle the argument and not jump to conclusions without all the data. What we may not know today we will eventually find out over time. Breakthrough infections (loosely used) occur but are pretty much uncommon overall. As news outlets focus on making headlines, scientists are conducting real-time studies to answer questions that continue to make news circles on a day today. As I referred to earlier, we have 3 studies that prove the effectiveness of the vaccines and how they prevent death and hospitalization. This is important because that has always been the focus of vaccine science since the Smallpox Epidemic of 1721 in Boston. The disease we know today as Covid-19 is a preventable disease, and science has proven that case time and time again. Unvaccinated people, as some scientists say, are on the clock, and with the emergence of a second wave on the rise, it may be a while before we can get this pandemic under control. However, for vaccinated individuals, the risk of infection is very present; continue to take precautions and study the science, not the social media memes.

Sources

- BC, Immunize. "What Are Vaccines?" Immunize BC, 19 May 2020, immunizebc.ca/what-are-vaccines.

- Berg, Sara. "What Doctors Wish Patients Knew about Breakthrough Covid Infections." American Medical Association, 6 Aug. 2021, www.ama-assn.org/delivering-care/public-health/what-doctors-wish-patients-knew-about-breakthrough-covid-infections.

- Bergwerk, Moriah, et al. "Covid-19 Breakthrough Infections in Vaccinated Health CARE WORKERS: NEJM." New England Journal of Medicine, 21 July 2021, www.nejm.org/doi/full/10.1056/NEJMoa2109072.

- Cavanaugh, Alyson M, et al. "Reduced Risk of REINFECTION with SARS-COV-2 AFTER COVID-19 Vaccination - KENTUCKY, May–June 2021." Centers for Disease Control and Prevention, Centers for Disease Control and Prevention, 6 Aug. 2021, www.cdc.gov/mmwr/volumes/70/wr/mm7032e1.htm.

- Chiwaya, Nigel, et al. "Data Shows How Rare SEVERE Breakthrough Covid Infections Are." NBCNews.com, NBCUniversal News Group, 6 Aug. 2021, www.nbcnews.com/specials/data-shows-how-rare-severe-breakthrough-covid-infections-are/index.html?utm_source=facebook&utm_medium=news_tab&utm_content=algorithm.

- Doll, Michelle. "Breakthrough Infections, Viral Load: What Does This Mean to You?" Breakthrough Infections Viral Load What Does This Mean to You | VCU Health, 2 Aug. 2021, www.vcuhealth.org/news/covid-19/breakthrough-infections-viral-load-what-does-this-mean-to-you.

- Dollomore, Doug. "MRNA Vaccines Slash Risk of COVID-19 Infection by 91 Percent in Fully Vaccinated People." University of Utah Health, University of Utah Health, 6 July 2021, healthcare.utah.edu/publicaffairs/news/2021/07/7-yoon-covid-vaccine.php.

- Fact Checker, Reuters. "Fact Check-Sars-Cov-2 Virus Began Mutating Prior to Mass Vaccine Rollouts." Reuters, Thomson Reuters, 23 July 2021, www.reuters.com/article/factcheck-mutations-vaccine-idUSL1N2OZ1PU.

- Fauzia, Miriam. "Fact Check: CDC Didn't Say COVID-19 Vaccinated Are 'Superspreaders', VACCINES FAILING." USA Today, Gannett Satellite Information Network, 7 Aug. 2021, www.usatoday.com/story/news/factcheck/2021/08/06/fact-check-cdc-didnt-say-covid-19-vaccinated-superspreaders/5475438001/.

- Kates, Jennifer, et al. "Covid-19 Vaccine Breakthrough Cases: Data from the States." KFF, 3 Aug. 2021, www.kff.org/policy-watch/covid-19-vaccine-breakthrough-cases-data-from-the-states/.

- McEvoy, Jemima. "Fully Vaccinated May Transmit Delta Just as Easily-and New Variant Shows Signs of Vaccine Evasion-Early U.k. Research Suggests." Forbes, Forbes Magazine, 6 Aug. 2021, www.forbes.com/sites/jemimamcevoy/2021/08/06/fully-vaccinated-may-transmit-delta-just-as-easily-and-new-variant-shows-signs-of-vaccine-evasion-early-uk-research-suggests/?utm_source=facebook&utm_medium=news_tab&utm_content=algorithm&sh=77c7e20a1ac5.

- Nania, Rachel. "8 Things to Know about BREAKTHROUGH COVID Infections." AARP, 6 Aug. 2021, www.aarp.org/health/conditions-treatments/info-2021/breakthrough-covid-infections.html.

- Respiratory Diseases, National Center for Immunization. "Understanding How Vaccines Work." Centers for Disease Control and Prevention, Centers for Disease Control and Prevention, 17 Aug. 2018, www.cdc.gov/vaccines/hcp/conversations/understanding-vacc-work.html.

- Sandoiu, Ana. "How Does the COVID-19 VACCINE Compare with Other EXISTING VACCINES?" Medical News Today, MediLexicon International, 14 Dec. 2020, www.medicalnewstoday.com/articles/how-do-covid-19-vaccines-compare-with-other-existing-vaccines.

- Walker, Molly. "Debate Is Over: COVID Vax Doubled Protection for the Previously Infected." Medical News, MedpageToday, 6 Aug. 2021, www.medpagetoday.com/infectiousdisease/covid19vaccine/93940.

- World Health, WHO. "Vaccine Efficacy, Effectiveness and Protection." World Health Organization, World Health Organization, 14 July 2021, www.who.int/news-room/feature-stories/detail/vaccine-efficacy-effectiveness-and-protection.

- Rafael Sanjuán, et al. "Viral Mutation Rates." Journal of Virology, 1 Oct. 2010, journals.asm.org/doi/10.1128/JVI.00694-10.

Sources

- Effort, Joint. "WHO-Convened Global Study of Origins of SARS-CoV-2: China Part." *World Health Organization*, World Health Organization, 10 Feb. 2021, www.who.int/publications/i/item/who-convened-global-study-of-origins-of-sars-cov-2-china-part.

- Biggers, Alana, and Tim Newman. "The Cell: Types, Functions, and Organelles." Medical News Today, MediLexicon International, 8 Feb. 2018, www.medicalnewstoday.com/articles/320878#division.

- Cann, Alan J. "Virus Mutation." Virus Mutation - an Overview | ScienceDirect Topics, 2012, www.sciencedirect.com/topics/biochemistry-genetics-and-molecular-biology/virus-mutation.

- Control, Centers Disease. "About Variants of the Virus That Causes COVID-19 ." Centers for Disease Control and Prevention, Centers for Disease Control and Prevention, 20 May 2021, www.cdc.gov/coronavirus/2019-ncov/variants/variant.html.

- Davis, Charles Patrick. "Medical Definition of DNA Virus." RxList, RxList, 29 Mar. 2021, www.rxlist.com/dna_virus/definition.htm.

- Davis, Charles Patrick. "Medical Definition of RNA Virus." MedicineNet, MedicineNet, 29 Mar. 2021, www.medicinenet.com/rna_virus/definition.htm.

- Davis, Charles Patrick. "Medical Definition of Zoonotic." MedicineNet, MedicineNet, 29 Mar. 2021, www.medicinenet.com/zoonotic/definition.htm.

- Davis, Charles Patrick. "What's a Virus? Viral Infection Types, Symptoms, Treatment." MedicineNet, MedicineNet, 6 Oct. 2020, www.medicinenet.com/viral_infections_pictures_slideshow/article.htm.

- Dhar, Michael. "What Is RNA?" LiveScience, Purch, 15 Oct. 2020, www.livescience.com/what-is-RNA.html.

- Enard, David, et al. "Viruses Are a Dominant Driver of Protein Adaptation in Mammals." ELife, ELife Sciences Publications, Ltd, 17 May 2016, elifesciences.org/articles/12469#digest.

- Flint, S. Jane, et al. Principles of Virology. Fifth ed., I, American Society for Microbiology, 2020.

- Gelderblom, Hans R. "Structure and Classification of Viruses." Medical Microbiology. 4th Edition., U.S. National Library of Medicine, 1 Jan. 1996, www.ncbi.nlm.nih.gov/books/NBK8174/#:~:text=A%20complete%20virus%20particle%20is,inside%20a%20symmetric%20protein%20capsid.

- Health, National Institute. "Coronaviruses." National Institute of Allergy and Infectious Diseases, U.S. Department of Health and Human Services, 26 Mar. 2021, www.niaid.nih.gov/diseases-conditions/coronaviruses.

- Health, National Institute. "What Is a Cell?: MedlinePlus Genetics." MedlinePlus, U.S. National Library of Medicine, 22 Feb. 2021, medlineplus.gov/genetics/understanding/basics/cell/#:~:text=Cells%20are%20the%20basic%20building%20blocks%20of%20all%20living%20things.&text=Cells%20also%20contain%20the%20body's,certain%20tasks%20within%20the%20cell.

- Health, National Institute. "What Is DNA?: MedlinePlus Genetics." MedlinePlus, U.S. National Library of Medicine, 19 Jan. 2021, medlineplus.gov/genetics/understanding/basics/dna/.

- Kuhn, Jens H., et al. "Virus Nomenclature below the Species Level: a Standardized Nomenclature for Natural Variants of Viruses Assigned to the Family Filoviridae." Archives of Virology, U.S. National Library of Medicine, 23 Sept. 2012, www.ncbi.nlm.nih.gov/pmc/articles/PMC3535543/#__ffn_sectitle.

- Racaniello, Vincent. "Vincent Racaniello." Virology Blog, 9 June 2004, www.virology.ws/2004/06/09/are-viruses-living/.

- Rampersad, Sephra, and Paula Tennant. "Replication and Expression Strategies of Viruses." Viruses, U.S. National Library of Medicine, 30 Mar. 2018, www.ncbi.nlm.nih.gov/pmc/articles/PMC7158166/#:~:text=All%20viruses%20must%20therefore%20express,vary%20among%20different%20viruses%20(Fig.

- Stern, Adi, and Raul Andino. "Viral Evolution: It Is All About Mutations." Viral Pathogenesis (Third Edition), Academic Press, 12 Feb. 2016, www.sciencedirect.com/science/article/pii/B9780128009642000173.

- Vidyasagar, Aparna. "What Are Viruses?" LiveScience, Purch, 6 Jan. 2016, www.livescience.com/53272-what-is-a-virus.html.

- Zoppi, Lois. "The Naming System Behind SARS-CoV-2." News, 9 Mar. 2021, www.news-medical.net/health/The-Naming-System-Behind-SARS-CoV-2.aspx.

- "Immune System (for Parents) - Nemours KidsHealth." Edited by Larissa Hirsch, KidsHealth, The Nemours Foundation, Oct. 2019, kidshealth.org/en/parents/immune.html.

- Davis, Charles P. "Medical Definition of Immune System." MedicineNet, MedicineNet, 29 Mar. 2021, www.medicinenet.com/immune_system/definition.htm.

- Health, National Institute. "NCI Dictionary of Cancer Terms." National Cancer Institute, 2021, www.cancer.gov/publications/dictionaries/cancer-terms/def/immune-system.

- Newman, Tim. "The Immune System: Cells, Tissues, Function, and Disease." Medical News Today, MediLexicon International, 11 Jan. 2018, www.medicalnewstoday.com/articles/320101#In-a-nutshell.

- Medicine, J H. "The Immune System." Johns Hopkins Medicine, John Hopkins Medicine, 2021, www.hopkinsmedicine.org/health/conditions-and-diseases/the-immune-system.

- Institue, National Health. "Overview of the Immune System." National Institute of Allergy and Infectious Diseases, U.S. Department of Health and Human Services, 2021, www.niaid.nih.gov/research/immune-system-overview.

- Institute, National Health. "NCI Dictionary of Cancer Terms." National Cancer Institute, 2021, www.cancer.gov/publications/dictionaries/cancer-terms/def/immunocompromised.

- Institue, National Health. "Immune Response: MedlinePlus Medical Encyclopedia." MedlinePlus, U.S. National Library of Medicine, 25 Mar. 2021, medlineplus.gov/ency/article/000821.htm#:~:text=Innate%2C%20or%20nonspecific%2C%20immunity%20is,materials%20from%20entering%20your%20body.

- Arizona, University of. Introduction to Immunology Tutorial, 24 Mar. 2000, www.biology.arizona.edu/immunology/tutorials/immunology/page3.html#:~:text=Adaptive%20immunity%20refers%20to%20antigen,designed%20to%20attack%20that%20antigen.

- Warrington R, Watson W, Kim HL, Antonetti FR. An introduction to immunology and immunopathology. Allergy Asthma Clin Immunol. 2011. doi:10.1186/1710-1492-7-S1-S1.

- Van Parijs L, Abbas AK. Homeostasis and self-tolerance in the immune system: turning lymphocytes off. Science. 1998;280(5361):243-248.

- Sakaguchi S. Naturally arising CD4+ regulatory T cells for immunologic self-tolerance and negative control of immune responses. Rev Immunol. 2004;22:531-562.

- Nikolich-Žugich J. Ageing and life-long maintenance of T-cell subsets in the face of latent persistent infections. Nat Rev Immunol. 2008;8(7):512-522.

- Health, Harvard. "How to Boost Your Immune System." *Harvard Health*, Harvard Health Publishing, 15 Feb. 2021, www.health.harvard.edu/staying-healthy/how-to-boost-your-immune-system.

- Ramesh, Sandhya, et al. "Immunity Boosters Are a Myth - Why You Shouldn't Believe Claims That Promise to Fight Covid." *ThePrint*, 4 Aug. 2020, theprint.in/health/immunity-boosters-are-a-myth-why-you-shouldnt-believe-claims-that-promise-to-fight-covid/470202/.

- Control & Prevention, Center Disease. "Symptoms of COVID-19." Centers for Disease Control and Prevention, Centers for Disease Control and Prevention, 2021, www.cdc.gov/coronavirus/2019-ncov/symptoms-testing/symptoms.html.

- Staff, Mayo Clinic. "Coronavirus Disease 2019 (COVID-19)." Mayo Clinic, Mayo Foundation for Medical Education and Research, 2 June 2021, www.mayoclinic.org/diseases-conditions/coronavirus/symptoms-causes/syc-20479963.

- David J Cennimo, MD. "Coronavirus Disease 2019 (COVID-19)." Practice Essentials, Background, Route of Transmission, Medscape, 20 May 2021, emedicine.medscape.com/article/2500114-overview.

- McIntosh, Kenneth. UpToDate, 9 June 2021, www.uptodate.com/contents/covid-19-epidemiology-virology-and-prevention#H2549483976.

- Caliendo, Angela, and Kimberly Hanson. Edited by Martin Hirsch and Allyson Bloom, UpToDate, UpToDate, 16 Apr. 2021, www.uptodate.com/contents/covid-19-diagnosis?topicRef=8298&source=see_link.

- Health, Minn Dept. "About COVID-19." Minnesota Dept. of Health, 8 June 2021, www.health.state.mn.us/diseases/coronavirus/basics.html.

- Division of Viral Diseases, National Center for Immunization and Respiratory Diseases (NCIRD). "Studying the Disease." Centers for Disease Control and Prevention, Centers for Disease Control and Prevention, 1 July 2020, www.cdc.gov/coronavirus/2019-ncov/science/about-epidemiology/studying-the-disease.html?CDC_AA_refVal=https%3A%2F%2Fwww.cdc.gov%2Fcoronavirus%2F2019-ncov%2Fcases-updates%2Fabout-epidemiology%2Fstudying-the-diesease.html.

- Nature, Journal of. Nature News, Nature Publishing Group, 2020, www.nature.com/subjects/viral-transmission.

- Cuffari, Benedette. "What Is Viral Shedding?" News, 16 Mar. 2021, www.news-medical.net/health/What-is-Viral-Shedding.aspx#:~:text=Throughout%20this%20ongoing%20process%2C%20infected,perform%20other%20normal%20daily%20activities.

- Health, National Institute. "Asymptomatic: MedlinePlus Medical Encyclopedia." MedlinePlus, U.S. National Library of Medicine, 2021, medlineplus.gov/ency/article/002217.htm.

- Meller, Megan. "The Asymptomatic and Pre-Symptomatic Spread of COVID-19." Gundersen Health System, 2020, www.gundersenhealth.org/covid19/the-asymptomatic-and-pre-symptomatic-spread-of-covid-19/.

- and Radiological Health, Center for Devices. "Antibody (Serology) Testing for COVID-19." U.S. Food and Drug Administration, FDA, 19 May 2021, www.fda.gov/medical-devices/coronavirus-covid-19-and-medical-devices/antibody-serology-testing-covid-19-information-patients-and-consumers.

- Marshall III, William M. "Here's What You Need to Know about COVID-19 Testing." Mayo Clinic, Mayo Foundation for Medical Education and Research, 12 Dec. 2020, www.mayoclinic.org/diseases-conditions/coronavirus/expert-answers/covid-antibody-tests/faq-20484429.

- Munster, Vincent J., et al. "A Novel Coronavirus Emerging in China - Key Questions for Impact Assessment: NEJM." New England Journal of Medicine, 21 Apr. 2021, www.nejm.org/doi/full/10.1056/NEJMp2000929?query=featured_coronavirus.

-

- Shimabukuro and Others, T.T., et al. "Importation and Human-to-Human Transmission of a Novel Coronavirus in Vietnam: NEJM." New England Journal of Medicine, 21 Apr. 2021, www.nejm.org/doi/full/10.1056/NEJMc2001272?query=featured_coronavirus.

- Yang, Qing, et al. Just 2% of SARS-CoV-2–Positive Individuals Carry 90% of the Virus Circulating in Communities, no. PNAS, 11 Apr. 2021, pp. 1–6., doi: https://doi.org/10.1073/pnas.2104547118.

- Newman, Tim. "Links between COVID-19 and Skin Rashes." Translated by Alexandra Saffins, Medical News Today, MediLexicon International, 18 Mar. 2021, www.medicalnewstoday.com/articles/links-between-covid-19-and-skin-rashes.

- Wyne, Kathleen. "Why Are People Developing Diabetes after Having COVID-19?" Ohio State Medical Center, 19 Feb. 2021, wexnermedical.osu.edu/blog/why-are-people-developing-diabetes-after-having-covid19.

- Clinic, Mayo. "COVID-19 (Coronavirus): Long-Term Effects." Mayo Clinic, Mayo Foundation for Medical Education and Research, 6 May 2021, www.mayoclinic.org/diseases-conditions/coronavirus/in-depth/coronavirus-long-term-effects/art-20490351#:~:text=Even%20in%20young%20people%2C%20COVID,Parkinson's%20disease%20and%20Alzheimer's%20disease.

- Division of Viral Diseases, National Center for Immunization and Respiratory Diseases (NCIRD). "Post-COVID Conditions." Centers for Disease Control and Prevention, Centers for Disease Control and Prevention, 8 Apr. 2021, www.cdc.gov/coronavirus/2019-ncov/long-term-effects.html.

- Health, UC Davis. "Long Haulers: Why Some People Experience Long-Term Coronavirus Symptoms." Long Haulers Suffer Long-Term Coronavirus Symptoms | UC Davis Health, 8 Feb. 2021, health.ucdavis.edu/coronavirus/covid-19-information/covid-19-long-haulers.html.

- Assoc, American Lung. "Learn about COVID-19." Learn about COVID-19 | American Lung Association, American Lung Association Scientific and Medical Editorial Review Panel., 14 Apr. 2021, www.lung.org/lung-health-diseases/lung-disease-lookup/covid-19/about-covid-19.

- Collins, Francis. "How Severe COVID-19 Can Tragically Lead to Lung Failure and Death." National Institutes of Health, U.S. Department of Health and Human Services, 10 May 2021, directorsblog.nih.gov/2021/05/11/how-severe-covid-19-can-tragically-lead-to-lung-failure-and-death/.

- Post, Wendy S. "Heart Problems after COVID-19." Johns Hopkins Medicine, 2021, www.hopkinsmedicine.org/health/conditions-and-diseases/coronavirus/heart-problems-after-covid19.

- Assoc, AAD. "COVID Toes, Rashes: How the Coronavirus Can Affect Your Skin." American Academy of Dermatology, 2021, www.aad.org/public/diseases/coronavirus/covid-toes.

- Bharat, MBBS, Prof Ankit, et al. Early Outcomes after Lung Transplantation for Severe COVID-19: a Series of the First Consecutive Cases from Four Countries, The Lancent, 31 Mar. 2021, www.thelancet.com/journals/lanres/article/PIIS2213-2600(21)00077-1/fulltext.

- Bharat, Ankit, et al. "Lung Transplantation for Patients with Severe COVID-19." Science Translational Medicine, American Association for the Advancement of Science, 16 Dec. 2020, stm.sciencemag.org/content/12/574/eabe4282.

- Weir, Kristen. "How COVID-19 Attacks the Brain." Monitor on Psychology, American Psychological Association, 11 Nov. 2020, www.apa.org/monitor/2020/11/attacks-brain#:~:text=In%20a%20review%20of%20case,in%20which%20the%20immune%20system.

- Hamilton, Jon. "How COVID-19 Attacks The Brain And May Cause Lasting Damage." NPR, NPR, 5 Jan. 2021, www.npr.org/sections/health-shots/2021/01/05/953705563/how-covid-19-attacks-the-brain-and-may-cause-lasting-damage.

- Shimabukuro and Others, T.T., et al. "Microvascular Injury in the Brains of Patients with Covid-19: NEJM." New England Journal of Medicine, 21 Apr. 2021, www.nejm.org/doi/full/10.1056/NEJMc2033369.

- University, Georgia State. "Study Finds COVID-19 Attack on Brain, Not Lungs, Triggers Severe Disease in Mice." ScienceDaily, ScienceDaily, 19 Jan. 2021, www.sciencedaily.com/releases/2021/01/210119114456.htm.

- Caffrey, Mary. "Viral RNA Found in Kidneys of COVID-19 Patients." AJMC, AJMC, 19 Aug. 2020, www.ajmc.com/view/lancet-covid-19-virus-can-attack-kidneys-speeding-death.

- Budson, Andrew. "What Is COVID-19 Brain Fog - and How Can You Clear It?" Harvard Health, 8 Mar. 2021, www.health.harvard.edu/blog/what-is-covid-19-brain-fog-and-how-can-you-clear-it-2021030822076.

- Fruend, Alexander. "How the Novel Coronavirus Attacks Our Entire Body: DW: 11.05.2020." DW.COM, Deutsche Welle, 5 Nov. 2020, www.dw.com/en/how-the-novel-coronavirus-attacks-our-entire-body/a-53389908.

- Madjid, Mohammad. "Potential Effects of Coronaviruses on the Cardiovascular System." JAMA Cardiology, JAMA Network, 1 July 2020, jamanetwork.com/journals/jamacardiology/fullarticle/2763846.

- Bleicher, Ariel, and Katherine Conrad. "We Thought It Was Just a Respiratory Virus." We Thought It Was Just a Respiratory Virus | UC San Francisco, University of California San Francisco, 10 June 2021, www.ucsf.edu/magazine/covid-body.

- Mallapaty, Smriti. "Mini Organs Reveal How the Coronavirus Ravages the Body." Nature News, Nature Publishing Group, 22 June 2020, www.nature.com/articles/d41586-020-01864-x.

- Marjot, Thomas, et al. "COVID-19 and Liver Disease: Mechanistic and Clinical Perspectives." Nature News, Nature Publishing Group, 10 Mar. 2021, www.nature.com/articles/s41575-021-00426-4.

- Pena, Luz. "SF Woman Who Will Have Fingers Amputated after Nearly Dying from COVID-19, Still Hesitant about Vaccine." ABC7 San Francisco, KGO-TV, 18 Dec. 2020, abc7news.com/coronavirus-latino-covid-rate-fingers-amputated-vaccine-news/8784805/.

- Martino, Giuseppe P, and Giuseppina Bitti. "DEFINE_ME." COVID Fingers: Another Severe Vascular Manifestation, European Society for Vascular Surgery. Published by Elsevier B.V. , 14 Aug. 2020, www.ejves.com/article/S1078-5884(20)30676-6/fulltext.

- Shimabukuro and Others, T.T., et al. "New-Onset Diabetes in Covid-19: NEJM." New England Journal of Medicine, 21 Apr. 2021, www.nejm.org/doi/full/10.1056/nejmc2018688.

- Lanese, Nicoletta. "COVID-19 May Trigger Diabetes in Some People." LiveScience, Purch, 22 Mar. 2021, www.livescience.com/covid19-may-trigger-diabetes.html.

- Accili, Domenico. "Can COVID-19 Cause Diabetes?" Nature News, Nature Publishing Group, 11 Jan. 2021, www.nature.com/articles/s42255-020-00339-7.

- Pawlowski, A. "'Covid Tongue' May Be Another Coronavirus Symptom, British Researcher Suggests." NBCNews.com, NBCUniversal News Group, 29 Jan. 2021, www.nbcnews.com/health/health-news/covid-tongue-may-be-another-coronavirus-symptom-british-researcher-suggests-n1256078.

- Pang, Wentai, et al. "Tongue Features of Patients with Coronavirus Disease 2019: a Retrospective Cross-Sectional Study." Integrative Medicine Research, Elsevier, 25 July 2020, www.sciencedirect.com/science/article/pii/S2213422020301256.

- Newman, Tim. "Links between COVID-19 and Skin Rashes." Translated by Alexandra Saffins, Medical News Today, MediLexicon International, 18 Mar. 2021, www.medicalnewstoday.com/articles/links-between-covid-19-and-skin-rashes.

- Wyne, Kathleen. "Why Are People Developing Diabetes after Having COVID-19?" Ohio State Medical Center, 19 Feb. 2021, wexnermedical.osu.edu/blog/why-are-people-developing-diabetes-after-having-covid19.

- Clinic, Mayo. "COVID-19 (Coronavirus): Long-Term Effects." Mayo Clinic, Mayo Foundation for Medical Education and Research, 6 May 2021, www.mayoclinic.org/diseases-conditions/coronavirus/in-depth/coronavirus-long-term-effects/art-20490351#:~:text=Even%20in%20young%20people%2C%20COVID,Parkinson's%20disease%20and%20Alzheimer's%20disease.

- Division of Viral Diseases, National Center for Immunization and Respiratory Diseases (NCIRD). "Post-COVID Conditions." Centers for Disease Control and Prevention, Centers for Disease Control and Prevention, 8 Apr. 2021, www.cdc.gov/coronavirus/2019-ncov/long-term-effects.html.

- Health, UC Davis. "Long Haulers: Why Some People Experience Long-Term Coronavirus Symptoms." Long Haulers Suffer Long-Term Coronavirus Symptoms | UC Davis Health, 8 Feb. 2021, health.ucdavis.edu/coronavirus/covid-19-information/covid-19-long-haulers.html.

- Assoc, American Lung. "Learn about COVID-19." Learn about COVID-19 | American Lung Association, American Lung Association Scientific and Medical Editorial Review Panel., 14 Apr. 2021, www.lung.org/lung-health-diseases/lung-disease-lookup/covid-19/about-covid-19.

- Collins, Francis. "How Severe COVID-19 Can Tragically Lead to Lung Failure and Death." National Institutes of Health, U.S. Department of Health and Human Services, 10 May 2021, directorsblog.nih.gov/2021/05/11/how-severe-covid-19-can-tragically-lead-to-lung-failure-and-death/.

- Post, Wendy S. "Heart Problems after COVID-19." Johns Hopkins Medicine, 2021, www.hopkinsmedicine.org/health/conditions-and-diseases/coronavirus/heart-problems-after-covid19.

- Assoc, AAD. "COVID Toes, Rashes: How the Coronavirus Can Affect Your Skin." American Academy of Dermatology, 2021, www.aad.org/public/diseases/coronavirus/covid-toes.

- Bharat, MBBS , Prof Ankit, et al. Early Outcomes after Lung Transplantation for Severe COVID-19: a Series of the First Consecutive Cases from Four Countries, The Lancent , 31 Mar. 2021, www.thelancet.com/journals/lanres/article/PIIS2213-2600(21)00077-1/fulltext.

- Bharat, Ankit, et al. "Lung Transplantation for Patients with Severe COVID-19." Science Translational Medicine, American Association for the Advancement of Science, 16 Dec. 2020, stm.sciencemag.org/content/12/574/eabe4282.

- Weir, Kristen. "How COVID-19 Attacks the Brain." Monitor on Psychology, American Psychological Association, 11 Nov. 2020, www.apa.org/monitor/2020/11/attacks-brain#:~:text=In%20a%20review%20of%20case,in%20which%20the%20immune%20system.

- Hamilton, Jon. "How COVID-19 Attacks The Brain And May Cause Lasting Damage." NPR, NPR, 5 Jan. 2021, www.npr.org/sections/health-shots/2021/01/05/953705563/how-covid-19-attacks-the-brain-and-may-cause-lasting-damage.

- Shimabukuro and Others, T.T., et al. "Microvascular Injury in the Brains of Patients with Covid-19: NEJM." New England Journal of Medicine, 21 Apr. 2021, www.nejm.org/doi/full/10.1056/NEJMc2033369.

- University , Georgia State. "Study Finds COVID-19 Attack on Brain, Not Lungs, Triggers Severe Disease in Mice." ScienceDaily, ScienceDaily, 19 Jan. 2021, www.sciencedaily.com/releases/2021/01/210119114456.htm.

- Caffrey, Mary. "Viral RNA Found in Kidneys of COVID-19 Patients." AJMC, AJMC, 19 Aug. 2020, www.ajmc.com/view/lancet-covid-19-virus-can-attack-kidneys-speeding-death.

- Budson, Andrew. "What Is COVID-19 Brain Fog - and How Can You Clear It?" Harvard Health, 8 Mar. 2021, www.health.harvard.edu/blog/what-is-covid-19-brain-fog-and-how-can-you-clear-it-2021030822076.

- Fruend, Alexander. "How the Novel Coronavirus Attacks Our Entire Body: DW: 11.05.2020." DW.COM, Deutsche Welle, 5 Nov. 2020, www.dw.com/en/how-the-novel-coronavirus-attacks-our-entire-body/a-53389908.

- Madjid, Mohammad. "Potential Effects of Coronaviruses on the Cardiovascular System." JAMA Cardiology, JAMA Network, 1 July 2020, jamanetwork.com/journals/jamacardiology/fullarticle/2763846.

- Bleicher, Ariel, and Katherine Conrad. "We Thought It Was Just a Respiratory Virus." We Thought It Was Just a Respiratory Virus | UC San Francisco, University of California San Francisco, 10 June 2021, www.ucsf.edu/magazine/covid-body.

- Mallapaty, Smriti. "Mini Organs Reveal How the Coronavirus Ravages the Body." Nature News, Nature Publishing Group, 22 June 2020, www.nature.com/articles/d41586-020-01864-x.

- Marjot, Thomas, et al. "COVID-19 and Liver Disease: Mechanistic and Clinical Perspectives." Nature News, Nature Publishing Group, 10 Mar. 2021, www.nature.com/articles/s41575-021-00426-4.

- Pena, Luz. "SF Woman Who Will Have Fingers Amputated after Nearly Dying from COVID-19, Still Hesitant about Vaccine." ABC7 San Francisco, KGO-TV, 18 Dec. 2020, abc7news.com/coronavirus-latino-covid-rate-fingers-amputated-vaccine-news/8784805/.

- Martino, Giuseppe P, and Giuseppina Bitti. "DEFINE_ME." COVID Fingers: Another Severe Vascular Manifestation, European Society for Vascular Surgery. Published by Elsevier B.V. , 14 Aug. 2020, www.ejves.com/article/S1078-5884(20)30676-6/fulltext.

- Shimabukuro and Others, T.T., et al. "New-Onset Diabetes in Covid-19: NEJM." New England Journal of Medicine, 21 Apr. 2021, www.nejm.org/doi/full/10.1056/nejmc2018688.

- Lanese, Nicoletta. "COVID-19 May Trigger Diabetes in Some People." LiveScience, Purch, 22 Mar. 2021, www.livescience.com/covid19-may-trigger-diabetes.html.

- Accili, Domenico. "Can COVID-19 Cause Diabetes?" Nature News, Nature Publishing Group, 11 Jan. 2021, www.nature.com/articles/s42255-020-00339-7.

- Pawlowski, A. "'Covid Tongue' May Be Another Coronavirus Symptom, British Researcher Suggests." NBCNews.com, NBCUniversal News Group, 29 Jan. 2021, www.nbcnews.com/health/health-news/covid-tongue-may-be-another-coronavirus-symptom-british-researcher-suggests-n1256078.

- Pang, Wentai, et al. "Tongue Features of Patients with Coronavirus Disease 2019: a Retrospective Cross-Sectional Study." Integrative Medicine Research, Elsevier, 25 July 2020, www.sciencedirect.com/science/article/pii/S2213422020301256.

- Affairs, Office of Regulatory. "Fraudulent COVID-19 Products." U.S. Food and Drug Administration, FDA, 10 June 2021, www.fda.gov/consumers/health-fraud-scams/fraudulent-coronavirus-disease-2019-covid-19-products.

- Ahmed, Irman. "The Disinformation Dozen: Center for Countering Digital Hate." CCDH, CCDH , 24 Mar. 2021, www.counterhate.com/disinformationdozen.

- Bill McCarthy, PolitiFact.com. "Fact-Check: Do COVID Vaccines 'Magnetize' People?" Statesman, Austin American-Statesman, 11 June 2021, www.statesman.com/story/news/politics/politifact/2021/06/11/sherri-tenpenny-makes-false-covid-vaccine-magnetism-claim-lawmakers/7653766002/.

- Caldera, Camille. "Fact Check: Operation Warp Speed Official Discussed Vaccine Distribution, Not Mandatory Vaccinations." USA Today, Gannett Satellite Information Network, 25 Nov. 2020, www.usatoday.com/story/news/factcheck/2020/11/24/fact-check-post-operation-warp-speed-official-missing-context/6398580002/.

- Caldera, Camille. "Fact Check: Strict Lockdowns, Experimental Vaccine Helped China Recover from COVID-19." USA Today, Gannett Satellite Information Network, 8 Dec. 2020, www.usatoday.com/story/news/factcheck/2020/12/04/fact-check-strict-lockdowns-covid-19-vaccine-helped-china-recover/3814394001/.

- Cao, Bin, et al. "A Trial of Lopinavir–Ritonavir in Adults Hospitalized with Severe Covid-19: NEJM." New England Journal of Medicine, 7 May 2020, www.nejm.org/doi/full/10.1056/NEJMoa2001282.

- Carballo-Carbajal, Iria. "No, Bill Gates Is Not Funding COVID-19 Vaccines as a Way to Conduct Global Surveillance or to Depopulate the World." Health Feedback, 8 July 2020, healthfeedback.org/claimreview/no-bill-gates-is-not-funding-covid-19-vaccines-as-a-way-to-conduct-global-surveillance-or-to-depopulate-the-world/.

- CC, MSK. "Fact Check: 7 Persistent Myths about COVID-19 Vaccines." Memorial Sloan Kettering Cancer Center, 24 May 2021, www.mskcc.org/coronavirus/myths-about-covid-19-vaccines.

- Clinic Staff, Mayo. "Vasovagal Syncope." Mayo Clinic, Mayo Foundation for Medical Education and Research, 19 Feb. 2021, www.mayoclinic.org/diseases-conditions/vasovagal-syncope/symptoms-causes/syc-20350527.

- Crist, Carolyn. "'Disinformation Dozen' Driving Anti-Vaccine Content." WebMD, WebMD, 25 Mar. 2021, www.webmd.com/children/vaccines/news/20210325/disinformation-dozen-driving-anti-vaccine-content.

- Dupuy, Beatrice. "Experts: MRNA Vaccine for COVID-19 Does Not Alter DNA." AP NEWS, Associated Press, 4 Sept. 2020, apnews.com/article/archive-fact-checking-9340521654.

- Fack Checker, Reuters. "Fact Check-Moderna's Chief Medical Officer Did Not Say MRNA Vaccines Alter DNA." Reuters, Thomson Reuters, 8 Apr. 2021, www.reuters.com/article/factcheck-moderna-mrna/fact-check-modernas-chief-medical-officer-did-not-say-mrna-vaccines-alter-dna-idUSL1N2M10IV.

- Fack Checkers, Reuters. "Fact Check: Video Makes Multiple False Claims about COVID-19 Pandemic." Reuters, Thomson Reuters, 3 Nov. 2020, www.reuters.com/article/uk-factcheck-pandemic-video/fact-check-video-makes-multiple-false-claims-about-covid-19-pandemic-idUSKBN27J2HM.

- Fact Check, Reuters. "Fact Check-No Evidence MRNA COVID-19 Vaccines Affect Sperm." Reuters, Thomson Reuters, 17 May 2021, www.reuters.com/article/factcheck-sperm-vaccine/fact-check-no-evidence-mrna-covid-19-vaccines-affect-sperm-idUSL2N2N42EC.

- Fact Checker, Reuters. "Fact Check: Anaphylaxis and Bell's Palsy Are Not the Most Common Side-Effects of COVID-19 Vaccines." Reuters, Thomson Reuters, 30 Jan. 2021, www.reuters.com/article/uk-factcheck-anaphylaxis/fact-check-anaphylaxis-and-bells-palsy-are-not-the-most-common-side-effects-of-covid-19-vaccines-idUSKBN29Z0PD.

- Fact Checker, Reuters. "Fact Check: Lipid Nanoparticles in a COVID-19 Vaccine Are There to Transport RNA Molecules." Reuters, Thomson Reuters, 5 Dec. 2020, www.reuters.com/article/uk-factcheck-vaccine-nanoparticles/fact-check-lipid-nanoparticles-in-a-covid-19-vaccine-are-there-to-transport-rna-molecules-idUSKBN28F0I9.

- Fact Checker, Reuters. "Partly False Claim: A List of Eight Coronavirus-Related 'Facts.'" Reuters, Thomson Reuters, 26 Mar. 2020, www.reuters.com/article/uk-factcheck-coronavirus-eight-facts/partly-false-claim-a-list-of-eight-coronavirus-related-facts-idUSKBN21D3EY.

- Funke, Daniel. "PolitiFact - Alternative Health Website Spreads False Claim about COVID-19 Vaccine Side Effects." @Politifact, 9 Dec. 2020, www.politifact.com/factchecks/2020/dec/09/blog-posting/alternative-health-website-spreads-false-claim-abo/.

- Gore, D'Angelo. "Hank Aaron's Death Attributed to Natural Causes." FactCheck.org, 28 Apr. 2021, www.factcheck.org/2021/01/scicheck-hank-aarons-death-attributed-to-natural-causes/.

- Greenberg, Jon. "PolitiFact - COVID-19 Skeptics Say There's an Overcount. Doctors in the Field Say the Opposite." @Politifact, 14 Apr. 2020, www.politifact.com/factchecks/2020/apr/14/candace-owens/covid-19-skeptics-say-theres-overcount-doctors-fie/.

- Gruber-Miller, Stephen. "Fact Check: Vitamins C and D Are Not Used in 'Conventional Treatment' of Coronavirus." USA Today, Gannett Satellite Information Network, 4 May 2020, www.usatoday.com/story/news/factcheck/2020/05/02/fact-check-coronavirus-covid-19-vitamins-c-d-treatment-joseph-mercola/3058491001/.

- Hui, Kayla. "CDC Study Confirms That COVID-19 Vaccines Block Transmission In the Real World." Verywell Health, 8 Apr. 2021, www.verywellhealth.com/cdc-study-covid-19-transmission-vaccines-5121080.

- Jones, Craig. "Claim That the FDA Found That Coronavirus Vaccines Awaiting Approval Could Cause Death Is Majorly Misleading." Newswise, 10 Dec. 2020, www.newswise.com/factcheck/claim-that-the-fda-found-that-coronavirus-vaccines-awaiting-approval-could-cause-death-is-majorly-misleading.

- Kancharla, Bharath, and Nanditha Kalidoss. "The Claims Made by Dr. Christiane Northrup Regarding COVID-19 Vaccines in This Video Are False." FACTLY, Factly, 20 Nov. 2020, factly.in/the-claims-made-by-dr-christiane-northrup-regarding-covid-19-vaccines-in-this-video-are-false/.

- Kiley, James P. "NIH Halts Clinical Trial of Hydroxychloroquine." National Institutes of Health, U.S. Department of Health and Human Services, 20 June 2020, www.nih.gov/news-events/news-releases/nih-halts-clinical-trial-hydroxychloroquine.

- Lovelace Jr, Berkeley, and Kevin Breuninger . "Trump Says He Takes Hydroxychloroquine to Prevent Coronavirus Infection Even Though It's an Unproven Treatment." CNBC, CNBC, 19 May 2020, www.cnbc.com/2020/05/18/trump-says-he-takes-hydroxychloroquine-to-prevent-coronavirus-infection.html.

- Mason, Jacquelyn. "The Nation of Islam and Anti-Vaccine Rhetoric." First Draft, 27 May 2021, firstdraftnews.org/articles/the-nation-of-islam-and-anti-vaccine-rhetoric/.

- McCarthy, Bill, et al. "Sherri Tenpenny." PolitiFact, 2021, www.politifact.com/personalities/sherri-tenpenny/.

- McLernon, Lianna Matt. "Zinc, Vitamin C Show No Effect for COVID-19 in Small Study." CIDRAP, 12 Feb. 2021, www.cidrap.umn.edu/news-perspective/2021/02/zinc-vitamin-c-show-no-effect-covid-19-small-study.

- Murai, Igor H, et al. "Effect of a Single High Dose of Vitamin D3 in Patients With Moderate to Severe COVID-19." JAMA, JAMA Network, 16 Mar. 2021, jamanetwork.com/journals/jama/fullarticle/2776738.

- Ngo, Madeleine. "Fact Check: Health and Human Services' Brett Giroir Confirms Vaccine Distribution Is Tracked to Ensure Dosing." USA Today, Gannett Satellite Information Network, 19 Jan. 2021, www.usatoday.com/story/news/factcheck/2021/01/19/fact-check-covid-19-vaccines-pfizer-moderna-brett-giroir-tracking/4207433001/.

- Pitts, D'Angela. "Here's Where That COVID-19 Vaccine Infertility Myth Came From-And Why It Is Not True." Henry Ford LiveWell, Henry Ford Health System Staff, 23 Apr. 2021, www.henryford.com/blog/2021/04/fertility-rumor-covid-vaccine.

- Press, Associated. "Reno Doctor's Selfie Used to Claim COVID-19 Is a Hoax | Health News | US News." U.S. News & World Report, U.S. News & World Report, 1 Dec. 2020, www.usnews.com/news/health-news/articles/2020-12-01/reno-doctors-selfie-used-to-claim-covid-19-is-a-hoax.

- Press, The Associated. "Post Makes False Claim about COVID-19 Vaccine Risk." AP NEWS, Associated Press, 1 Feb. 2021, apnews.com/article/fact-checking-afs:Content:9934822788.

- Press, The Associated. "Research Shows COVID-19 Was Not Manufactured in a Lab." AP NEWS, Associated Press, 16 Sept. 2020, apnews.com/article/archive-fact-checking-9391149002.

- Rougerie, Pablo. "No Aborted Fetal Tissue or Cells in the Johnson & Johnson COVID-19 Vaccine." Health Feedback, 25 Mar. 2021, healthfeedback.org/claimreview/no-aborted-fetal-tissue-or-cells-in-the-johnson-johnson-covid-19-vaccine/.

- Rourke, Ciara. "PolitiFact - People Are Dying from COVID-19." @Politifact, 4 Sept. 2020, www.politifact.com/factchecks/2020/sep/04/viral-image/people-are-dying-covid-19/.

- Sadeghi, McKenzie. "Fact Check: Posts Falsely Claim Cap Was on Nancy Pelosi's COVID-19 Vaccine." USA Today, Gannett Satellite Information Network, 19 Dec. 2020, www.usatoday.com/story/news/factcheck/2020/12/19/fact-check-photos-show-nancy-pelosi-receiving-covid-19-vaccine/3973360001/.

- Self, Wesley H, et al. "Effect of Hydroxychloroquine on Clinical Status at 14 Days in Hospitalized Patients With COVID-19." JAMA, JAMA Network, 1 Dec. 2020, jamanetwork.com/journals/jama/fullarticle/2772922.

- Skeptic, Original. "Kelly Brogan Archives." Skeptical Raptor, 2 May 2021, www.skepticalraptor.com/skepticalraptorblog.php/tag/kelly-brogan/.

- Smith, Michelle R, and Johnatan Reiss. "How Vaccine Disinformation Super Spreaders Have Cashed in on Americans' Fears during Pandemic." Chicagotribune.com, 13 May 2021, www.chicagotribune.com/coronavirus/vaccine/ct-aud-nw-vaccine-disinformation-20210513-7lid6y5jlzakdjow3wfzwwxruy-story.html.

- Spencer , Saranac Hale, and Angelo Fichera. "RFK Jr. Video Pushes Known Vaccine Misrepresentations." FactCheck.org, 28 Apr. 2021, www.factcheck.org/2021/03/scicheck-rfk-jr-video-pushes-known-vaccine-misrepresentations/.

- Staff, Reuters. "Fact Check: RFID Microchips Will Not Be Injected with the COVID-19 Vaccine, Altered Video Features Bill and Melinda Gates and Jack Ma." Reuters, Thomson Reuters, 4 Dec. 2020, www.reuters.com/article/uk-factcheck-vaccine-microchip-gates-ma/fact-check-rfid-microchips-will-not-be-injected-with-the-covid-19-vaccine-altered-video-features-bill-and-melinda-gates-and-jack-ma-idUSKBN28E286.

- Strauss, Valerie. "Analysis | Debunking Anti-Vaxxer RFK Jr.'s Claim about 'Suspicious' Coronavirus Vaccine Deaths, a Phony Elon Musk Tweet and More News Literacy Lessons." The Washington Post, WP Company, 5 Feb. 2021, www.washingtonpost.com/education/2021/02/05/news-literacy-refuting-rfkjr-phony-elon-musk-tweet/.

- Tech, Fora. "In Spite of Evidence to the Contrary, Rizza Islam Claims That the MMR Vaccine Is Used for Depopulation." Science Feedback, Health Feedback, 19 Dec. 2019, sciencefeedback.co/claimreview/in-spite-of-evidence-to-the-contrary-rizza-islam-claims-that-the-mmr-vaccine-is-used-for-depopulation/.

- Teoh, Flora. "Frequency of Deaths in Elderly Individuals after COVID-19 Vaccination Wasn't Higher than the Frequency in Those Who Weren't Vaccinated." Health Feedback, 27 Jan. 2021, healthfeedback.org/claimreview/frequency-of-deaths-in-elderly-individuals-after-covid-19-vaccination-werent-higher-than-the-frequency-in-those-who-werent-vaccinated/.

- Teoh, Flora. "The U.S. National Childhood Vaccine Injury Act Does Not Stop People from Suing Vaccine Manufacturers." Health Feedback, 7 Apr. 2021, healthfeedback.org/claimreview/the-u-s-national-childhood-vaccine-injury-act-does-not-stop-people-from-suing-vaccine-manufacturers/.

- Thomas, Suma, et al. "Effect of Zinc and Ascorbic Acid on Symptom Length Among Patients With SARS-CoV-2." JAMA Network Open, JAMA Network, 12 Feb. 2021, jamanetwork.com/journals/jamanetworkopen/fullarticle/2776305.

- UC Davis Health, Public Affairs and Marketing. "UC Davis Experts: Science Says Wearing Masks and Social Distancing Slow COVID-19 (VIDEO)." UC Davis Health, 6 July 2020, health.ucdavis.edu/health-news/newsroom/uc-davis-experts-science-says-wearing-masks-and-social-distancing-slow-covid-19/2020/07.

- USA, EPA. "Protect Your Family from Sources of Lead." EPA, Environmental Protection Agency, 22 Dec. 2020, www.epa.gov/lead/protect-your-family-sources-lead#:~:text=Deteriorating%20lead%2Dbased%20paint%20(peeling,Doors%20and%20door%20frames%3B%20and.

- VĒBERE, ILZE. "Diskreditēts Osteopāts Izplata Melus Par Covid-19 Pandēmiju." A Discredited Osteopath Spreads Lies about the Covid-19 Pandemic, 26 May 2020, rebaltica.lv/2020/05/vai-tiesam-viens-cilveks-fauci-vainojams-covid-19-pandemija/.

- Zubeyir, Ayşe E. "Kendini Doktor Olarak Tanıtan Ben Tapper'ın PCR Testiyle Ilgili Iddiaları: Teyit." Edited by Sosyal M Medya, Şüpheli Bilgileri Inceleyen Doğrulama Platformu, Teyit, 22 Dec. 2020, teyit.org/analiz-dr-ben-tapperin-pcr-testine-yonelik-iddialari.

- Zhou, Hong, et al. *Redirecting*, Cell.com, 3 June 2021, doi.org/10.1016/j.cell.2021.06.008.

- Flint, S. Jane, et al. *Principles of Virology*. Fifth ed., II, Wiley/ASM Press, 2020.

- Zhou, Hong, et al. "Identification of Novel Bat Coronaviruses Sheds Light on the Evolutionary Origins of SARS-CoV-2 and Related Viruses." Cell, Cell Press, 9 June 2021, www.sciencedirect.com/science/article/pii/S0092867421007091?via%3Dihub.

- Andersen, Kristian G., et al. "The Proximal Origin of SARS-CoV-2." Nature News, Nature Publishing Group, 17 Mar. 2020, www.nature.com/articles/s41591-020-0820-9#Sec8.

- Sandoiu, Ana. "The New Coronavirus Was Not Man-Made, Study Shows." Medical News Today, MediLexicon International, 20 Mar. 2020, www.medicalnewstoday.com/articles/the-new-coronavirus-was-not-genetically-engineered-study-shows#Ending-the-rumors-about-SARS-CoV-2-.

- Wessner, David R. Nature News, Nature Publishing Group, 2010, www.nature.com/scitable/topicpage/the-origins-of-viruses-14398218/.

- BC, Immunize. "What Are Vaccines?" Immunize BC, 19 May 2020, immunizebc.ca/what-are-vaccines.

- Berg, Sara. "What Doctors Wish Patients Knew about Breakthrough Covid Infections." American Medical Association, 6 Aug. 2021, www.ama-assn.org/delivering-care/public-health/what-doctors-wish-patients-knew-about-breakthrough-covid-infections.

- Bergwerk, Moriah, et al. "Covid-19 Breakthrough Infections in Vaccinated Health CARE WORKERS: NEJM." New England Journal of Medicine, 21 July 2021, www.nejm.org/doi/full/10.1056/NEJMoa2109072.

- Cavanaugh, Alyson M, et al. "Reduced Risk of REINFECTION with SARS-COV-2 AFTER COVID-19 Vaccination - KENTUCKY, May–June 2021." Centers for Disease Control and Prevention, Centers for Disease Control and Prevention, 6 Aug. 2021, www.cdc.gov/mmwr/volumes/70/wr/mm7032e1.htm.

- Chiwaya, Nigel, et al. "Data Shows How Rare SEVERE Breakthrough Covid Infections Are." NBCNews.com, NBCUniversal News Group, 6 Aug. 2021, www.nbcnews.com/specials/data-shows-how-rare-severe-breakthrough-covid-infections-are/index.html?utm_source=facebook&utm_medium=news_tab&utm_content=algorithm.

- Doll, Michelle. "Breakthrough Infections, Viral Load: What Does This Mean to You?" Breakthrough Infections Viral Load What Does This Mean to You | VCU Health, 2 Aug. 2021, www.vcuhealth.org/news/covid-19/breakthrough-infections-viral-load-what-does-this-mean-to-you.

- Dollomore, Doug. "MRNA Vaccines Slash Risk of COVID-19 Infection by 91 Percent in Fully Vaccinated People." University of Utah Health, University of Utah Health, 6 July 2021, healthcare.utah.edu/publicaffairs/news/2021/07/7-yoon-covid-vaccine.php.

- Fact Checker, Reuters. "Fact Check-Sars-Cov-2 Virus Began Mutating Prior to Mass Vaccine Rollouts." Reuters, Thomson Reuters, 23 July 2021, www.reuters.com/article/factcheck-mutations-vaccine-idUSL1N2OZ1PU.

- Fauzia, Miriam. "Fact Check: CDC Didn't Say COVID-19 Vaccinated Are 'Superspreaders', VACCINES FAILING." USA Today, Gannett Satellite Information Network, 7 Aug. 2021, www.usatoday.com/story/news/factcheck/2021/08/06/fact-check-cdc-didnt-say-covid-19-vaccinated-superspreaders/5475438001/.

- Kates, Jennifer, et al. "Covid-19 Vaccine Breakthrough Cases: Data from the States." KFF, 3 Aug. 2021, www.kff.org/policy-watch/covid-19-vaccine-breakthrough-cases-data-from-the-states/.

- McEvoy, Jemima. "Fully Vaccinated May Transmit Delta Just as Easily-and New Variant Shows Signs of Vaccine Evasion-Early U.k. Research Suggests." Forbes, Forbes Magazine, 6 Aug. 2021, www.forbes.com/sites/jemimamcevoy/2021/08/06/fully-vaccinated-may-transmit-delta-just-as-easily-and-new-variant-shows-signs-of-vaccine-evasion-early-uk-research-suggests/?utm_source=facebook&utm_medium=news_tab&utm_content=algorithm&sh=77c7e20a1ac5.

- Nania, Rachel. "8 Things to Know about BREAKTHROUGH COVID Infections." AARP, 6 Aug. 2021, www.aarp.org/health/conditions-treatments/info-2021/breakthrough-covid-infections.html.

- Respiratory Diseases, National Center for Immunization. "Understanding How Vaccines Work." Centers for Disease Control and Prevention, Centers for Disease Control and Prevention, 17 Aug. 2018, www.cdc.gov/vaccines/hcp/conversations/understanding-vacc-work.html.

- Sandoiu, Ana. "How Does the COVID-19 VACCINE Compare with Other EXISTING VACCINES?" Medical News Today, MediLexicon International, 14 Dec. 2020, www.medicalnewstoday.com/articles/how-do-covid-19-vaccines-compare-with-other-existing-vaccines.

- Walker, Molly. "Debate Is Over: COVID Vax Doubled Protection for the Previously Infected." Medical News, MedpageToday, 6 Aug. 2021, www.medpagetoday.com/infectiousdisease/covid19vaccine/93940.

- World Health, WHO. "Vaccine Efficacy, Effectiveness and Protection." World Health Organization, World Health Organization, 14 July 2021, www.who.int/news-room/feature-stories/detail/vaccine-efficacy-effectiveness-and-protection.

- Rafael Sanjuán, et al. "Viral Mutation Rates." Journal of Virology, 1 Oct. 2010, journals.asm.org/doi/10.1128/JVI.00694-10.

- Farry, R F, et al. "Early Appearance of Two Distinct GENOMIC Lineages OF SARS-CoV-2 in Different Wuhan Wildlife Markets Suggests SARS-CoV-2 Has a Natural Origin." *Virological*, 12 May 2021, virological.org/t/early-appearance-of-two-distinct-genomic-lineages-of-sars-cov-2-in-different-wuhan-wildlife-markets-suggests-sars-cov-2-has-a-natural-origin/691.

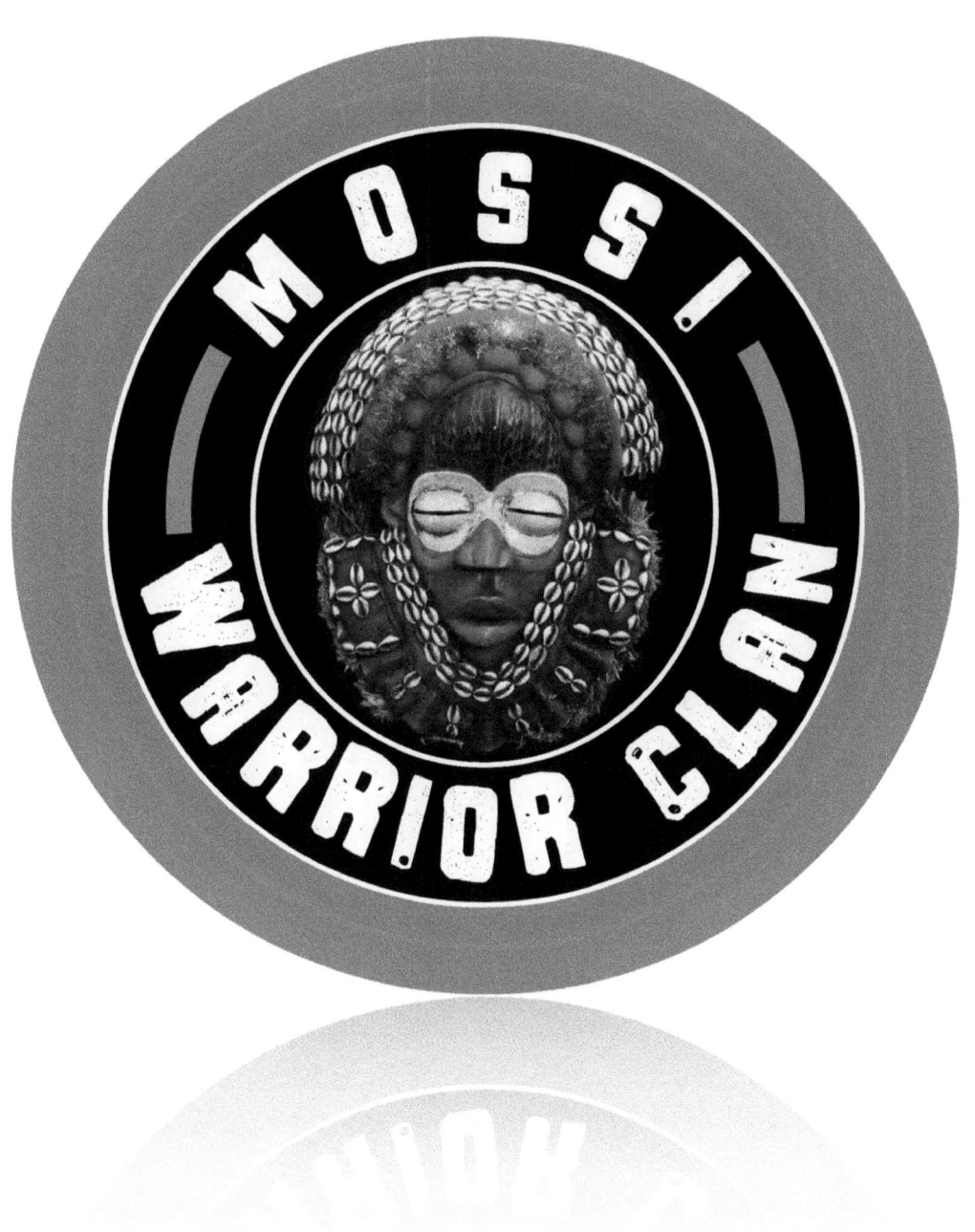

Mossi Warrior Clan Presents

From Cocaine to
Consciousness

By Ini-Herit Khalfani

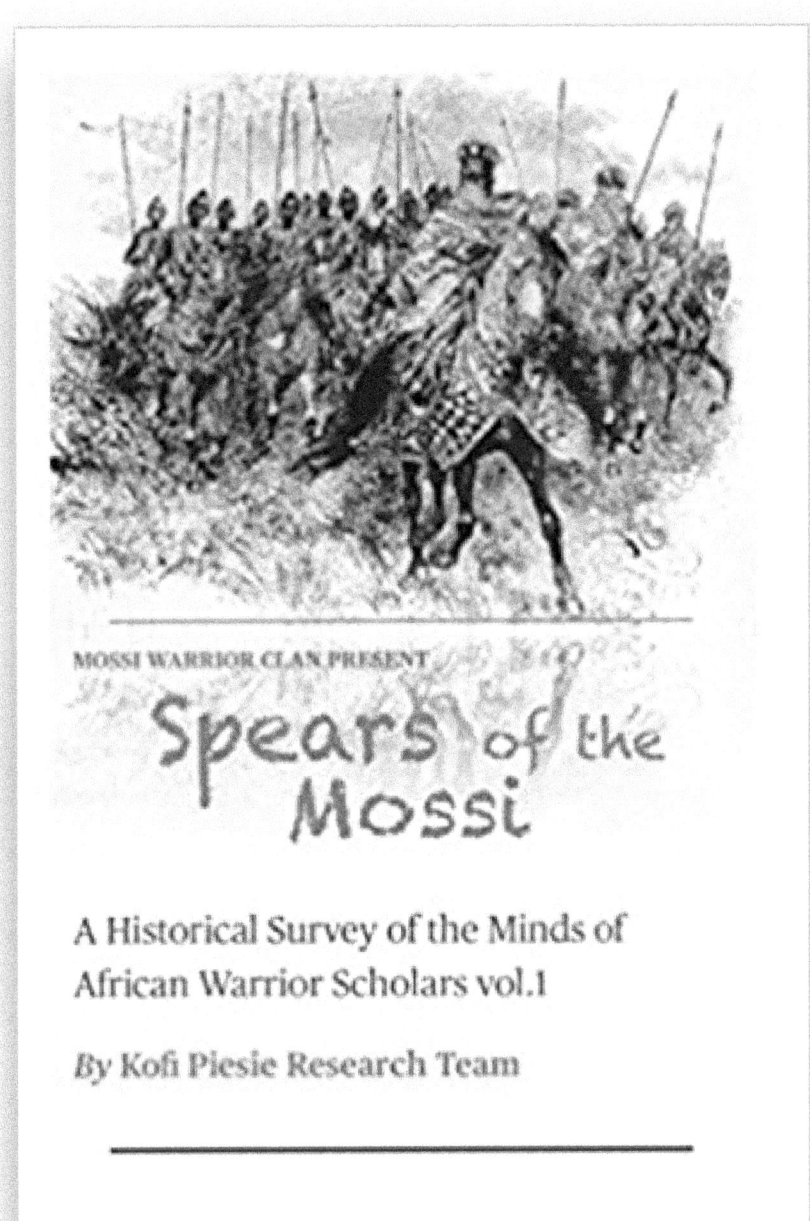

MOSSI WARRIOR CLAN PRESENT

Spears of the Mossi

A Historical Survey of the Minds of African Warrior Scholars vol.1

By Kofi Piesie Research Team

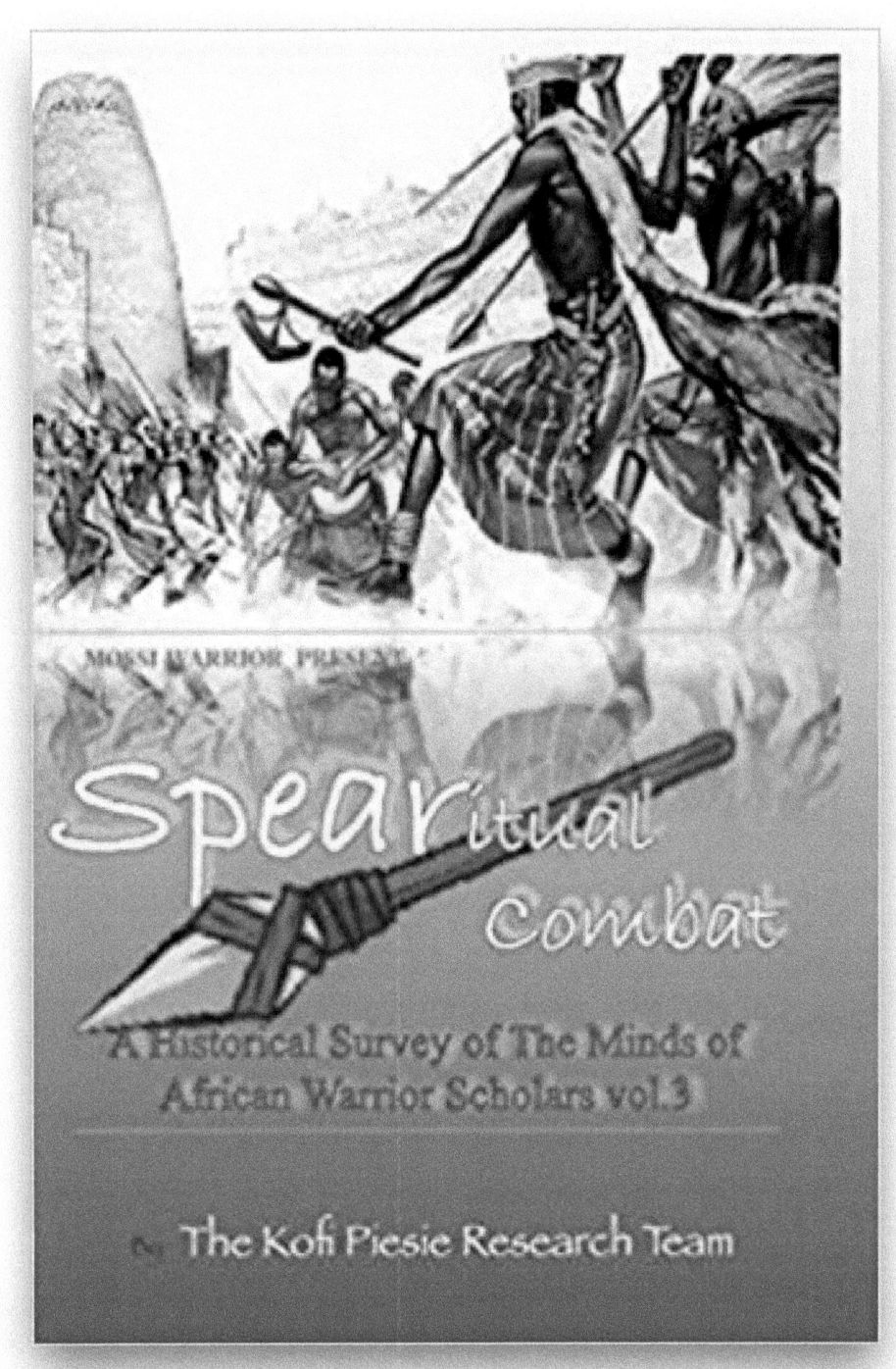

www.ingramcontent.com/pod-product-compliance
Ingram Content Group UK Ltd.
Pitfield, Milton Keynes, MK11 3LW, UK
UKHW051659240426
12048UKWH00039B/1500